PENNSYLVANIA

Trolleys

In Color

VOLUME III

The Pittsburgh Region

West Penn Railways
Johnstown Traction Company
Pittsburgh Railways Company

(Edward S. Miller)

By William D. Volkmer

RWY. COMPANY		TIME	
		1	15
THIS TRANSFER Acceptable only as Designated on the Reverse Side within Time Limit or first connection there-after on Date of Issuance.		30	45
		2	15
		30	45
		3	15
		30	45
FEATURING:		**4**	15
		30	45
		5	15
WEST PENN RAILWAYS COMPANY		30	45
		6	15
		30	45
		7	15
		30	45
JOHNSTOWN TRACTION COMPANY		**8**	15
		30	45
		9	15
		30	45
PITTSBURGH RAILWAYS COMPANY		**10**	15
		30	45
		11	15
		30	45
		12	15
A. M.	IF COUPON IS DETACHED	30	45

050198

P.M.

VOID IF DETACHED

Published by
Morning Sun Books, Inc.

9 Pheasant Lane
Scotch Plains, NJ 07076

Library of Congress
Catalog Card No. 97-070598

First Printing
ISBN 1-58248-019-2

Color separation and printing by
The Kutztown Publishing Co., Inc.
Kutztown, Pennsylvania

DEDICATION

This book is dedicated to my mother, Madeline Doehrel Volkmer. It was she who took me on my first trolley ride, when I was 10 years old, from Wollaston, Mass. to downtown Boston. That was in 1946. Today, 53 years later, I realize that ride must have made quite an impression on me, for this is my third book on the subject of trolleys.

ACKNOWLEDGEMENTS

Without the immense help of Edward H. Lybarger of the Pennsylvania Trolley Museum, the section on West Penn in this volume would not be nearly as complete as it is. The photo collections of Eugene Van Dusen, photos taken by James P. Shuman and Edward S. Miller augment those taken by the author to round out the illustration of this book. Many of these photos, particularly those taken by Ed Miller, have previously been published in black and white in the book West Penn Railways by the Pennsylvania Railway Museum Association as well as the Central Electric Railfans Association book on the same subject. For more details on this fascinating line, readers are encouraged to obtain a copy of those books. The PRMA book was originally published in 1973 and reprinted in 1986. My sincere thanks go to all who have helped. As in the previous volume, Bob Abrams, Art Ellis, and Sam James proofread the manuscript, as did Ed Lybarger, but the inevitable mistakes can be credited solely to the author. The West Penn map and some of the timetables are courtesy of the Miller Library of the Pennsylvania Trolley Museum. Sam James contributed additional paper to round out the book. As always, Bob Yanosey, Gail Gottlund, and the graphics artists at Kutztown Publishing particularly, have contributed to the overall pleasing appearance of the book.

PENNSYLVANIA Trolleys

In Color

VOLUME III

The Pittsburgh Region

West Penn Railways
Johnstown Traction Company
Pittsburgh Railways Company

Volume III of *Pennsylvania Trolleys In Color* completes our color photographic tour of Pennsylvania. In this, and the previous two volumes, we visited all of the properties operating during the era of color photography, and here we shall cover the three western Pennsylvania operations that existed post-1940.

West Penn was by far the least photographed in color, because it expired in 1952, prior to the general advent of the single-lens reflex camera, and probably equally as important, the general inaccessibility of the line by paralleling roads. While it is true that the Pennsylvania Turnpike gave fans speedy access to Western Pennsylvania, once there, most photography was either done in the cities, at passing sidings, or on fan trips.

Pittsburgh Railways in the 40s, 50s and 60s represented the last remaining mega-trolley system in the nation. During the decades of the 40s and 50s, trolleys accounted for the lion's share of transit vehicles over a system that could only be described

as enormous. PRCo.'s 666 car PCC fleet was surpassed only by Chicago, a system much smaller geographically, but more densely populated. In the 1960s we used to quip that Pittsburgh Railways operated streetcars on lines that other cities would not even offer *transit* service, the patronage was so light and the headways so poor.

In 1960, Johnstown Traction Company represented the very last small city trolley operation in the nation. The severe depression of 1958 dwindled patronage to a mere trickle and the city became a veritable Mecca for trolley fans. Your author once chartered two different cars for a three-hour tour of the system, splitting the $27 cost with one other railfan, for a "party of two charter!" Naturally, under conditions such as those, we were allowed to motor the car, while the assigned motorman, the railfan favorite, Johnny Yuchek, sat in the front seat and observed, or supervised.

Follow along with us now, as we photographically tour Western Pennsylvania and its trolley treasures, in color.

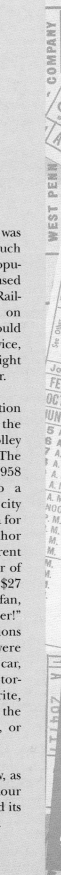

West Penn Railways Company
Connellsville, Pennsylvania

Operationally headquartered in Connellsville, PA, (Main Offices were at 14 Wood Street, Pittsburgh) the West Penn Railways Company was the result of the merger of as many as 34 small independent local trolley operations in the bituminous coal and coke producing region east of Pittsburgh. The utility company owner, West Penn Electric Company, also owned the trolley companies in Wheeling, West Virginia, as well as interurban routes out of Fairmont and Clarksburg, West Virginia. The West Virginia interurbans were standard gauge lines, known as Monongahela - West Penn Public Service, Inc.

Lines of the West Penn System connected McKeesport, Trafford, Greensburg, Scottdale, Connellsville and Uniontown. Branches also operated to Latrobe, Fairchance, Martin, and Brownsville. Lines unconnected to the main operation ran between Oakdale-McDonald, New Kensington-Aspinwall, Leechburg-Apollo, and Kittanning-Ford City.

By 1940, the weaker lines had been abandoned, including local streetcar lines in McKeesport and Greensburg, the Allegheny Valley Route in New Kensington, the Leechburg-Apollo line, the route between Kittanning and Ford City, plus the "Hunker line" between Greensburg and Scottdale.

The principal car fleet remaining in 1940 consisted of about 68 cars. All but two were converted by 1945/46 to one-man operation. There were about twenty-eight front-entrance, center-exit interurban cars, numbered in the 700-series. These had been built between 1910 and 1925 by Cincinnati Car Company and the West Penn's own company shops in Connellsville. During World War II, nine more 700s were returned to service, newly converted to one-man operation.

For lighter lines and local service, there were a dozen cars built between 1921 and 1928, again by Cincinnati Car Company and West Penn, numbered in the 280 and 290-series. Two veteran survivors of the early era were cars 204 and 212, built in 1901

by Stephenson. The 212 was used only on tripper runs in the latter days. A few other seldom used 200-series arch roof and 600-series deck roofed cars rounded out the West Penn wartime roster.

Following the abandonment of the Allegheny Valley Route between Aspinwall and New Kensington in 1937, the 12 modern (1929) Cincinnati curved side cars (831-842) were brought down to the Coke Region, where they too performed local service on the Fairchance, Latrobe, Phillips, and South Connellsville lines.

Freight service was provided throughout the system using only the company's express motors, as a result of the broad (5' 2½") track gauge. Daily through freight over the Pittsburgh Railways trackage from Trafford to downtown Pittsburgh was provided until September 15, 1941. Because this operation was largely a nighttime affair and was discontinued early in the era of color photography, few black and white photos are known to exist and no color photos. There was one day trip in each direction, but few if any railfans seem to have been able to catch them on film.

Beginning on August 15, 1942, service was abandoned from Larimer to Trafford, followed by further cutbacks from Larimer to Irwin, on January 3, 1948. Large-scale abandonments came soon after, beginning in 1950 with Uniontown-Brownsville service, cut on January 28th, followed by Uniontown-Martin a week later, on February 4th, and Uniontown-Fairchance on March 25th.

On January 20, 1951 the lines from Connellsville to Dickerson Run and Connellsville-Uniontown via Phillips were abandoned.

The summer of 1952 saw the total demise of the system. Irwin to Greensburg was discontinued July 12th, Hecla Junction to Latrobe on August 2nd, and the final abandonment, Greensburg to Uniontown via the Main Line, plus the South Connellsville local line, August 9, 1952.

Ridership over the routes was almost entirely local in nature, with very few riders traveling the entire distance between the major cities and towns.

The downfall of the rail system coincided with the general decline and obsolescence of the soft coal industry in the region. The coal mines had become exhausted and tightening of pollution laws even in those early days hastened their closure. One of the major customers of the coal mined here was the Pennsylvania Railroad, which was at the time converting from steam to diesel power. Most coke was now produced in modern by-product ovens located near the steel mills.

Because the area contained few paralleling roads, the substitute bus service was poor and West Penn discontinued bus service on June 20, 1953. One lone trolley locomotive continued to serve the shop at Connellsville until 1958, at which time it was donated to the Pennsylvania Trolley Museum at Arden, PA. The museum also has in its possession car 832, the bodies of 722 and 739, plus Monongahela-West Penn Baldwin steeple cab locomotive 3000 as well as MWP combine 274.

Scheduled speeds were quite slow as a result of the frequent stops, and service was usually "by the clock", either hourly, half hourly, or two hourly.

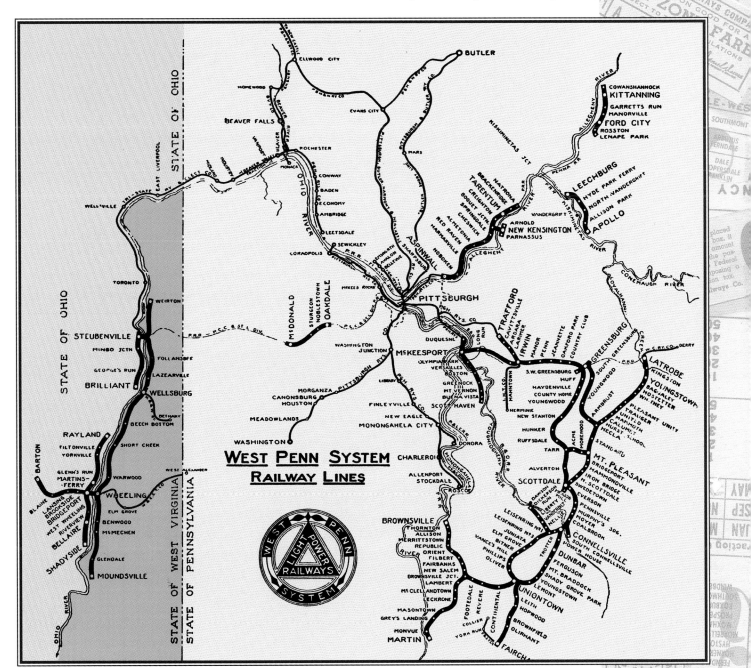

Map courtesy of Pennsylvania Trolley Museum. Only the routes shown in the lower right hand quadrant existed during the color photography era and hence, are the lines covered in this book.

CONNELLSVILLE

Connellsville was the hub of activities on the far flung West Penn system. The company shops were located here and lines radiated out in four directions. One might say that Connellsville was the last great interurban "hub" left in America in 1952 when the system was abandoned. Lines operated from here to Greensburg, two routes to Uniontown, a line to Dickerson Run, plus a short city type line to South Connellsville.

Passengers enter both front and rear doors as they load up at the Connellsville terminal on the last day of service in Connellsville, August 9, 1952. The bus on the adjacent track takes passengers to Dawson (Dickerson Run), a point not reached by the cars since January 20, 1951. (**Edward S. Miller**)

Photographer Miller climbed all those steps on West Fairview Street in Connellsville to catch the 294 making yet another "last day run" to South Connellsville. The track in the foreground was for freight cars and laid up passenger cars at the terminal. (**Edward S. Miller**)

Here we see the car turning the corner from West Fairview onto South Arch Street for the run to South Connellsville. The building in the background is the impressive Connellsville terminal. Cars bound for Uniontown and Greensburg turned right at this point and turned again at the extreme left of the picture. After this date, Saturday August 9, 1952, the only cars to make this turn would be the farewell fantrip the next day. **(Edward S. Miller)**

The 2½ mile line to South Connellsville was one of only two West Penn routes originally constructed to standard gauge (Kittanning was the other). West Penn acquired the line in 1902 and converted the track to 5' 2½" gauge a year later. The route was strictly streetcar in nature, serving among others, a glass plant at South Connellsville. It was one of the last lines to operate on the system, being abandoned on August 9, 1952. This line was generally home to the only remaining intact West Penn car (number 832) now being restored at the Pennsylvania Trolley Museum near Washington, PA. Two other carbodies are being reclaimed, but will have trucks other than the original installed under them when restored.

On the last day of revenue passenger service, car 294 picks up one of the regular riders at Woodlawn Avenue and Race Street in South Connellsville. The date is Saturday August 9, 1952. The two-car South Connellsville line met all interurbans in downtown Connellsville and passed each other at Race Street siding, just ahead of the 294. **(Edward S. Miller)**

THE COMPANY SHOPS, CONNELLSVILLE

West Penn maintained carbarns at Latrobe, South Greensburg, Uniontown and small storage barns near Jeannette and Scottdale, but the preponderance of car maintenance as well as car construction occurred at the main Connellsville Shops, located on the Southwest side of town, adjacent to the Pennsy tracks. It was separated from the rest of the system at that point by the elevated Western Maryland Railroad tracks. Close by was Leisenring Junction, where the interurban lines to Uniontown via Philips and via the Main Line diverged. New car construction ceased at the shops in the late 1920s and all remaining cars purchased were from Cincinnati Car Company.

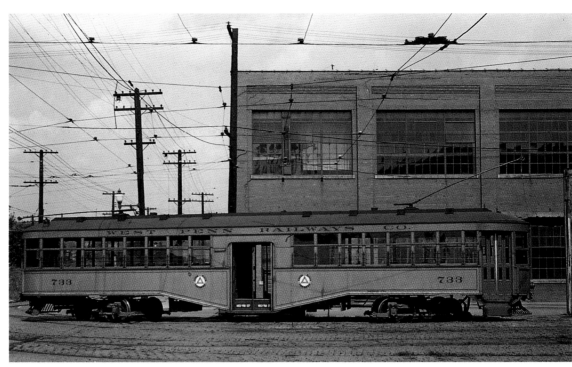

Car 733 was captured on color film outside the Connellsville shop on August 31, 1941. It would survive the war years and participate in the last day ceremonies almost eleven years later. **(James P. Shuman)**

The last operable piece of equipment on the West Penn system was this diminutive locomotive, a 1915 product of the Connellsville Shops. After this photo was taken on February 10, 1952, loco number 1 soldiered on at the shop that was taken over by the power company until 1958 when it was donated to the Pennsylvania Trolley Museum, where it remains to this day. The unit was originally built to standard gauge so as to run on Pennsy industrial siding tracks, but it was soon regauged to broad gauge and the short distance on the Pennsy was rebuilt to dual gauge. **(Edward S. Miller)**

Curved side car 833 sits outside of the Connellsville Shop building on August 31, 1941, about two years after it was brought down from the Allegheny Valley route in New Kensington for further service on the main part of the West Penn System. These cars were rumored to have been plagued with motor problems, but it is thought that the cars were simply too small to handle wartime crowds, hence the reactivation of surplus 700-series cars during the war. **(James P. Shuman)**

LEISENRING JUNCTION

The intersection of Eighth Street and Leisenring in Connellsville made for some interesting trackwork. The special work was basically two single tracks crossing at 90 degrees with connecting switches at three of the four quadrants. The four quadrants were, north to Connellsville and Greensburg, east to Connellsville Shop, south to Uniontown via main line, and west to Uniontown via Philips and Dawson via Vanderbuilt Junction. The two photos on this page depict the area surrounding this point.

West Penn 712 has just departed the Connellsville Shops and is ducking under the Western Maryland Railroad tracks on Leisenring Street. Just ahead of the car is the three-way junction at Eighth Street known as Leisenring Junction. The date is May 29, 1949. The WM 4-6-6-4, the railroad track and the entire West Penn have long since ceased to be part of the landscape at this location. The one exception to this is that the shops are now visible from this vantage point because of the railroad embankment removal!
(James P. Shuman)

Car 703 is southbound out of Connellsville on 8th Street and is turning right towards Uniontown via the Phillips line, on May 29, 1949. The motorman is in his usual standing up position with his right hand on the brake handle, his left hand on the controller, and his eyes on the switchwork, ever wary of an impending derailment. The 703 was the oldest 700-series car in service at the time, but it became a victim of cannibalization prior to the last year of service, in order to keep the other cars operating. **(James P. Shuman)**

On May 25, 1952, a southbound car to Connellsville from Greensburg operates on Broadway in Scottdale. The main offices of the H.C. Frick Coke Company are on the right. At the far end of the street the car will turn left onto the bridge over the PRR and B&O railroads which was shared with U.S. Route 119 into the town of Everson. **(Eugene Van Dusen Collection)**

Northbound car 726 tiptoes across the U.S. 119 bridge en route from Brown Street, Everson to South Broadway and Newman Streets in Scottdale. The date was Friday, August 8, 1952. Southbound automobiles were obligated to pass to the left of the oncoming trolley and run the risk of a head-on collision with an opposing automobile! **(Edward S. Miller)**

Buttermore siding was near the end of a private right-of-way for southbound cars heading into downtown Connellsville. From here on in, Nachod signals governed the movements over city streets. In the accompanying photo the distant pole (just to the right of the front of the car) is adorned by one of these instruments.

Fading into the sunset north of Connellsville, the 719 parallels U.S. 119 for the long descent into town. On Friday evening August 8, 1952, the sun was fast setting on the entire West Penn network of lines, for the next day would be the last to have cars burnishing these rails. **(Edward S. Miller)**

On Sunday afternoon, August 10, 1952 all is quiet at Connellsville shops as car 294 and others have had their trolley poles pulled for the last time. None of this series cars were saved. **(Edward S. Miller)**

HUFF CARHOUSE, SOUTH GREENSBURG

West Penn maintained a small carhouse in South Greensburg along Huff Avenue. The two views on this page depict cars laying over at that location, somewhat overshadowed by mine tailings nearby that belonged to the Keystone Coal and Coke Company, Greensburg #1 mine. This barn was situated alongside the former Hunker Line that was abandoned in 1939.

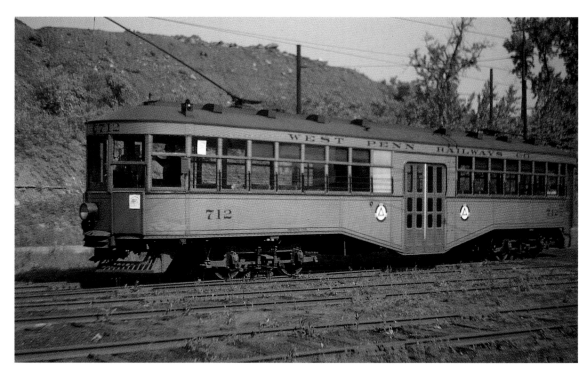

*Car 712 lays over at Greensburg carhouse on May 29, 1949. (**James P. Shuman**)*

*Car 291 is at the same location, May 24, 1949. This car would become one of the last three cars to ever operate on the system. (**James P. Shuman**)*

Greensburg served as a hub of sorts with West Penn cars converging there from points as far away as Trafford, Irwin, McKeesport, and Jeannette to the west, and Connellsville from the south. A fair sized terminal graced the downtown area with a small freight depot behind it. Service ended on the local streetcar lines in the Greensburg area in 1937 and from Irwin on July 12, 1952.

A northbound car approaches Sand House siding south of South Greensburg on May 29, 1949. The road in the foreground is appropriately named "Broadway". About 3/10ths of a mile ahead, the car will enter street trackage at Broad and Reamer to continue on into downtown Greensburg some two miles distant. Fifty years later, this scene (minus the tracks) remains relatively unchanged. (**James P. Shuman**)

Car 724 approaches downtown Greensburg on Mt. Pleasant Street at the intersection with Main Street, August 8, 1952. The overcast in the sky adds to the gloom of the impending trolley abandonment the next day. The trackage on Mt. Pleasant Street was added in 1936 just before the local line Highland Avenue/ Pittsburgh and Otterman Streets was abandoned. (**Edward S. Miller**)

GREENSBURG TERMINAL

Local car to Irwin, the 287 is seen loading at the usual point on the north side of the Greensburg terminal, May 29, 1949. The advertised "frequent service" is represented by the next 700-series interurban car to Uniontown, already waiting in the wings on the loop behind the terminal, in the company of the ubiquitous Ford transit bus of the era. The Irwin car will turn left on the switch in the foreground, the Uniontown car will turn right. **(James P. Shuman)**

Car 712, obviously a fantrip car because it is operating in the wrong direction on the Greensburg station loop, sits in front of the unused West Penn Greensburg freight house on May 29, 1949. Last freight service on West Penn was operated in 1941. **(James P. Shuman)**

The fantrip motorman has brought the 712 around to the sunny side of the Greensburg Terminal for a pose prior to departure for Connellsville on May 29, 1949. **(James P. Shuman)**

The rear of the beautiful Greensburg station is obscured by cars 739 and 733 as they round the loop one last time, August 10, 1952. The building now serves as the Greensburg City Hall. **(Edward S. Miller)**

WEST PENN RAILWAYS

SCHEDULE

Greensburg—Jeannette—Irwin
January 26, 1948

GREENSBURG

For Jeannette, Penn, Manor, Irwin

A. M.—5.00, 5.20 to Penn, 5.30, 6.00, 6.30, 7.00, 7.30, 8.00, 8.30, 9.00, 9.30, 10.00, 10.30, 11.00, 11.30, 12.00.

P. M.—12.30, 1.00, 1.30, 2.00, 2.30, 3.00, 3.30, 4.00, 4.30, 5.00, 5.30, 6.00, 6.30, 7.00, 7.30, 8.00, 8.30, 9.00, 9.30, 10.00, 10.30, 11.00, 11.30, to Penn.

5.00, 5.20, and 5.30 A. M. cars do not run Sunday.

IRWIN

For Manor, Penn, Jeannette and Greensburg.

WEEK DAYS

A. M.—6.00, 6.30, 7.00, 7.30, 8.00, 8.30, 9.00, 9.30, 10.00, 10.30, 11.00, 11.30, 12.00.

P. M.—12.30, 1.00, 1.30, 2.00, 2.30, 3.00, 3.50, 4.20, 4.50, 5.20, 5.50, 6.10, 6.30, 7.00, 7.30, 8.00, 8.30, 9.00, 9.30, 10.00, 10.30, 11.00, 11.30, 12.00.

SATURDAY

A. M.—6.00, 6.30, 7.00, 7.30, 8.00, 8.30, 9.00, 9.30, 10.00, 10.30, 11.00, 11.30, 12.00.

P. M.—12.30, 1.00, 1.40, 2.20, 2.50, 3.20, 3.50, 4.20, 4.50, 5.20, 5.50, 6.20, 6.50, 7.20, 7.50, 8.20, 8.50, 9.10, 9.30, 10.00, 10.30, 11.00, 11.30, 12.00.

SUNDAY

A. M.—7.00, 7.30, 8.00, 8.30, 9.00, 9.30, 10.00, 10.30, 11.00, 11.30, 12.00.

P. M.—12.30, 1.00, 1.30, 2.00, 2.30, 3.00, 3.50, 4.00, 4.30, 5.00, 5.30, 6.00, 6.30, 7.00, 7.50, 8.00, 8.30, 9.00, 9.30, 10.00, 10.30, 11.00, 11.30, 12.00.

Greensburg-Irwin Line

The line to Irwin from Greensburg, during the 1940s, was actually a truncation of the former Trafford line. Poor patronage on the Irwin to Trafford segment led to its earlier abandonment. Service between Greensburg and Trafford was operated almost exclusively with the 280-290-series cars. Prior to 1940, low 200 series cars served this route. The McKeesport cars were pressed into service after the 800s failed to perform in snow on this hilly route with a lot of paved tracks. Connections were made at Trafford with the Pittsburgh Railways, although few people used the trolleys all the way to downtown Pittsburgh. The Pennsy enjoyed almost all of that business. The Trafford line was cut back first to Larimer on August 14, 1942, then to Irwin on January 3, 1948. Final day for the line was July 12, 1952.

The end of the Irwin line after it was cut back to that point was at this location. Behind the 288 was a trestle that led to Larimer and Trafford. The July 4, 1948 photo was taken at the corner of 4th & Main Streets in Irwin. (**Eugene Van Dusen Collection**)

Two cars pass at Penn Siding alongside the highway between Penn and Manor, PA, about a mile and a half short of Manor, on May 31, 1952 just prior to abandonment of the Irwin line. The weed-infested right-of-way was indicative of the deferred maintenance West Penn was subjected to during its final year or two of operation. The highway was later to become designated as State Route 4010.
(**James P. Shuman**)

The heaviest route for West Penn operated from Greensburg on the north, past Hecla Junction, Mount Pleasant, Connellsville and Dunbar, to Uniontown, a distance of 30.6 miles, taking 2 hours and 25 minutes. Just about the only persons to make the entire trip were rail-fans. An alternate line from Greensburg to Uniontown operated via Hunker and rejoined the Main Line at Scottdale. It was abandoned in 1939, whereas the Main Line was retained until the end of all service, August 9, 1952.

The three principal cities, Greensburg, Connellsville and Uniontown each had very impressive downtown terminal stations that they shared with intercity bus operators.

Sweeper number 2, formerly stationed at Greensburg, passed northbound car 724 at Iron Bridge Siding on August 8, 1952, making one last trip to Connellsville shops where it will be scrapped. Sweeper number 4, the Latrobe assignment had made it to Connellsville by August 3rd. From the looks of the right-of-way, the corn will be growing in the cinders come next summer! **(Edward S. Miller)**

Northbound car 709 treads cautiously over Blue Ridge crossing below Murphy Siding, (below Pennsville and above Moyer, PA), Friday August 8, 1952. Only one West Penn crossing was known to have been protected by wig-wag signals, Poplar Grove, just north of Connellsville. **(Edward S. Miller)**

At Scottdale, the line crossed over the railroad tracks on this bridge. Car 732 was going over on August 8, 1952. A view of a car on the bridge from street level appears in the section on Scottdale. **(Edward S. Miller)**

On Friday, August 8, 1952, northbound car 733 is seen crossing the bridge above Reed Station in Armbrust, PA, just north of Hecla Junction en route to Greensburg. **(Edward S. Miller)**

For many years the author drove the Pennsylvania Turnpike between New Stanton and Breezewood, keeping an eye out for the West Penn bridge, just to see it was still in place. This was the main line crossing overhead between Hecla Junction and Scottdale. The Turnpike was constructed in 1939-40 and opened for business on October 1, 1940. This, of course was after the 1939 abandonment of the Hunker line through New Stanton. The photos on this page show the two fantrip cars making one last trip over the Turnpike, on August 10, 1952.

The trademark cinder ballasted West Penn right-of-way is evident in this photo as we see the farewell fantrip cars posing on the bridge over the Pennsylvania Turnpike. The cinders will offer little resistance to foliage incursion after this date as the cars, rails and tracks will be no more. **(Edward S. Miller)**

After shooting the above photo, photographer Miller moved over to get a view of the overpass for posterity. The girders were left standing after this August 10, 1952 last run and remained intact until they were removed in June 1968. **(Edward S. Miller)**

HECLA JUNCTION

The main line between Greensburg and Connellsville was joined at Hecla Junction by a branch from Latrobe. For many years, a small deli store stood in the center of the wye and doubled as a waiting room. A month or so before abandonment, "Wally's" was torn down and a small waiting shelter substituted.

Car 739 passes Hecla Junction southbound on a fantrip special, the day following abandonment of all remaining service. Another special car is seen on the bridge carrying the branch to Latrobe.
(Edward S. Miller)

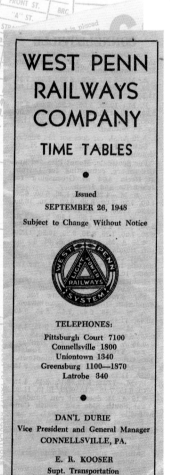

WEST PENN
RAILWAYS
COMPANY

TIME TABLES

•

Issued
SEPTEMBER 26, 1948

Subject to Change Without Notice

TELEPHONES:
Pittsburgh Court 7100
Connellsville 1800
Uniontown 1340
Greensburg 1100—1870
Latrobe 340

•

DAN'L DURIE
Vice President and General Manager
CONNELLSVILLE, PA.

E. R. KOOSER
Supt. Transportation
CONNELLSVILLE, PA.

A view looking northeast at the same location on the same day shows cars 739 and 733.
(Edward S. Miller)

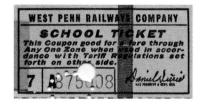

Another shot of Hecla Junction with car 739 on August 10, 1952 shows the fact that the ubiquitous deli/store in the center of the wye has been removed, probably in anticipation of the line's impending demise. A waiting shelter replaced it. (**Edward S. Miller**)

In this photo, car 733 is seen crossing the bridge at Hecla Junction on the segment leading to Latrobe via Calumet, Whitney and Baggaley. (**Edward S. Miller**)

MOUNT PLEASANT

Mount Pleasant was situated south of Hecla Junction and north of Scottdale and Connellsville. Prior to September 16, 1936 there was a line that connected Mt. Pleasant with Tarr, a town on the Greensburg-Con-nellsville "Hunker Line". A lone car shuttled back and forth once an hour between the two towns, connecting the two lines before it was discontinued. The Hunker line lasted three years longer before it too was abandoned, on June 3, 1939.

Prior to the abandon-ment of the Hunker route, all of the Latrobe cars operated south of Hecla Junction to Mt. Pleasant to make the connection. After 1939, Hecla Junction was the end of the Latrobe route.

The same photo stop as above, but looking north from in front of the fire station. The return trip later on from Greensburg would mark the last visit by a trolley to Mt. Pleasant.
(Edward S. Miller)

Car 739 leads a two car Pittsburgh Electric Railway Club special on the farewell trip, northbound, on August 10, 1952. The car is at the corner of W. Main Street and Center. The classic Pennsylvania brick street paving is evident in this photo. **(Edward S. Miller)**

Calumet was a small mining community made up of company houses known in the Western Pennsylvania vernacular as a "patch". It was located about two miles up the Latrobe line from Hecla Junction. A PRR spur line served the mines in the area.

Car 733, the PERC post-abandonment day special car, has operated from Hecla Junction as far up the Latrobe line as powered overhead would permit. A breaker in the middle of Calumet bridge prevented further operation up the line that had been abandoned a week earlier. The "patch" can be seen under the bridge structure.
(Edward S. Miller)

The two car special is seen out on the bridge prior to turning back towards Hecla Junction on the abandoned Latrobe Line August 10, 1952. **(Edward S. Miller)**

LATROBE LINE

The Latrobe line was essentially a shuttle line operating between the town of Latrobe, on the Pennsy main line, south to Hecla Junction, a distance of 13.8 miles. At Hecla, passengers transferred to main line Greensburg-Uniontown cars. Cars operated on an hourly headway, but were spliced by short turns to Baggaley, located a little over 4 miles south of Latrobe. An interurban line, operating in 1952 making 4-mile short turns, was indicative of the localized nature that characterized West Penn ridership. Begun as a local streetcar line in 1899, it was purchased by West Penn in 1904, extended to Hecla Junction in 1914, and abandoned on August 2, 1952.

Just over two miles south of the Latrobe car house lay Kingston Siding. In the photo (above) two cars pass at Kingston siding on the line from Latrobe to Hecla Junction. The Hecla cars operated on an hourly headway. The northbound car on the right was one of the Baggaley trippers. (**James P. Shuman**)

A northbound car approaches Kingston siding. A few hundred feet south of here the car had crossed Loyalhanna Creek on a trestle. The rolling terrain covered by West Penn presented a bit of a challenge to the magnetic track brakes on the West Penn cars. It was because of these grades that WP chose to adopt electric brakes in 1905. More specifically, the grade on E. Crawford Avenue in Connellsville was the determining factor in the company's decision to forsake the air brake. Only the 800-series Cincinnati cars utilized air brakes during the later days of system operation. Both photos taken May 31, 1947.
(**James P. Shuman**)

Baggaley served as a short turn point for tripper cars out of Latrobe. South of Baggaley, the terrain consisted of gently rolling countryside devoted chiefly to farming after the area's mines played out.

This pastoral scene about a half mile south of Baggaley has changed over the years as the farms have disappeared along with the orange interurbans. Heavy vegetation obscures the view today. The coal mine at Baggaley was closed in 1922 and the nearby Hostetter mine was also long gone by 1947. The stone house on the right has since had an addition put on it. The photo was taken May 31, 1947. Information is courtesy of Edward Lybarger who revisited the site in 1998. (James P. Shuman)

WHITNEY SIDING

Whitney siding was the only "left handed" siding on the Latrobe line. Here railfan extra car 712 meets the regular 713 on what apparently was a chilly May 29, 1947. This siding was installed in 1944 to replace one that formerly existed one-half mile west (timetable south). The red light illuminated on the roof of the 713 indicates to the opposing motorman that no cars are following in his block.
(James P. Shuman)

BRIDGES AND TRESTLES

Because of the hilly terrain that the West Penn Railways traversed, not to mention the plethora of railroads and mine spurs, there were numerous trestles and bridges along the route. These structures tended to lend a sense of identity and charm to the property and were a veritable magnet for photographers eager to compose interesting photographs of the line. Pictured here is a sampling of those structures, all now long obliterated from the Pennsylvania landscape.

Car 729 is crossing over the Western Maryland Bowest Yard where the WM interchanged with the B&O. The main line in the foreground is the B&O Sheepskin Branch from Connellsville to Uniontown and Fairmont, West Virginia. This shot was also made Saturday afternoon, August 9, 1952. (**Edward S. Miller**)

*A view looking up at the Calumet Bridge on the Latrobe line shows the two farewell fantrip cars on August 10, 1952. The Latrobe line had been abandoned the previous week so this point was as far as the cars could operate that day. (**Eugene Van Dusen Collection**)*

*Leaving Brownsville Junction for Martin, car 720 crosses the Footedale trestle outbound on a sunny February 5, 1950, the last time a car would visit the line, as it was abandoned to service the previous day. (**James P. Shuman**)*

Iron Bridge

One of the more photographed trestles along the line was at Iron Bridge. Located north of Scottdale on the Main Line it traversed the Pennsylvania Railroad Southwest Branch that ran from Greensburg to Fairchance. There was a carhouse a short distance from this bridge that was the original Mt. Pleasant Division home, but after 1911 it served only as a storage facility for unneeded cars.

On Friday, August 8, 1952, car 732 makes a northbound crossing over the trestle at Iron Bridge. The track curving off to the left of the photo belongs to the B&O Broadford to Mt. Pleasant branch serving the Standard Works. Views look west. **(Both, Edward S. Miller)**

A southbound car operates alongside the B&O and PRR south of Dunbar, PA on the last day of regular service, August 9, 1952. The photo was taken from a railroad overpass that entered the Dunbar Furnace Company facility. Gist Run Creek babbles merrily alongside the car.
(James P. Shuman)

Car 725 crosses northbound over the PRR overpass north of Fayette Siding and north of Dunbar. The date was August 9, 1952, one of the last regularly scheduled cars over the line. Three railroads were crossed in this general area. This and the B&O stayed the same for all 49 years of West Penn existence. The Western Maryland crossing area just south of here changed at least twice over the years.
(Both, Edward S. Miller)

PENNSYLVANIA TROLLEYS
ONE FARE
West Penn Railways Company

THE PHILLIPS LINE

Connellsville was connected to Uniontown by two separate lines, again showing the localized nature of West Penn ridership. Hourly service was provided between the two towns via Leisenring and Phillips, covering the 13.9 miles in 50 minutes. The main line cars covered 13.5 miles between the same two towns taking 58 minutes, indicating a heavier ridership and/or more stops.

On the half-hour, cars operated out of Connellsville as far as Leisenring Junction and turned right towards Dickerson Run. Both lines were abandoned January 20, 1951.

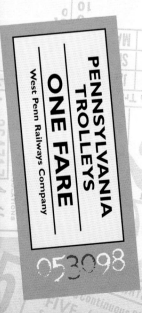

A southbound car on the Phillips line crosses a trestle at Vance's Mill on March 19, 1950. The Vance's Mill Branch of the Pennsylvania Railroad passed under the trestle. That line served West Leisenring, also known as Bute, and sometimes referred to as Leisenring #2. (**James P. Shuman**)

UNIONTOWN

A car leaves Uniontown climbing the trestle over the Pennsylvania Railroad towards Jordan wye located at the north end of the bridge. At the wye, northbound cars turned right towards Connellsville (via the Phillips line) and southbound cars turned left towards Brownsville and Martin. Photo taken June 1, 1947. *(James P. Shuman)*

Prior to August 1913, West Penn Railways had an on-street terminal on West Main Street. Heavy traffic congestion was being experienced at that point. In 1913 a new off-street terminal was built and shortly thereafter, in 1914, a second line was opened from Uniontown to Connellsville, via Phillips. Cars entered town via a long trestle shown in the photo left. In 1930, a wye, known as Jordan wye was constructed in order to allow cars to and from Brownsville and Martin to avoid trackage on Main Street altogether and share trackage with the Phillips cars into the Uniontown terminal. At this time a second new terminal was constructed with a freight house nearby. That terminal remained until the end of service in August 1952.

Blue Ridge Bus Lines, a subsidiary of Potomac Edison Company and thus a cousin of West Penn Railways, shared the Uniontown terminal with the trolleys. Here car 291 makes one last loop around the station during the post-abandonment fantrip on August 10, 1952. The building henceforth is truly a "bus terminal." The station was on Penn Street near Beeson Avenue and the B&O yard. The view looks north towards the rear of the complex. (Edward S. Miller)

FAIRCHANCE LINE

It took thirty minutes for cars to make the 7.2-mile run from Uniontown to Fairchance. Cars ran half-hourly except on Sunday, when they were hourly. Service was discontinued March 25, 1950, ending all West Penn operation south of Uniontown.

Fairchance bound 719 rounds the corner from Stewart Avenue onto Fayette Street in Uniontown on March 19, 1950. The line would be abandoned a week later, with the last cars operating March 25th. With the disappearance of the Martin and Brownsville lines two months earlier, abandonment of the Fairchance line meant that only the lines north of Uniontown remained in service. (**James P. Shuman**)

At the end of the Fairchance line car 706 has changed ends for a return run to Uniontown on March 6, 1950. (**Eugene Van Dusen Collection**)

This is the end of the line in more ways than one. Car 720 rests at the end of the Martin line on February 5, 1950, the day following cessation of regular service. The trip was a special charter to make one last visit to the southernmost point on the West Penn System. **(Eugene Van Dusen Collection)**

Spliced between the Brownsville cars out of Uniontown, the Martin line cars branched left at Brownsville Junction, 7.1 miles out of Uniontown, traversed rolling countryside, through McClelland-town and Masontown ending just north of Martin. This line, at 19.7 miles, was 3 miles longer than Browns-ville, making it diffi-cult to maintain the same headways as the Brownsville cars. Thus, on Saturdays, when riding was par-ticularly heavy, a shuttle car, usually the 212, was pressed into service between Masontown and Martin. It also oper-ated on the Browns-ville line on Saturdays. The line was aban-doned one week after the Brownsville line, on February 4, 1950.

Car 720 stops at Brownsville Junction en route to Martin on the farewell fantrip, Sunday February 5, 1950. Visible in this photo are the two hand thrown switches that the motorman actuated to signal his entrance into the next single-track block. The lights on the right hand side of the track were lit protecting his car from behind and warning opposing cars of his presence in the block. The other switch killed the lights in the previous block indicating he was in the clear. **(Eugene Van Dusen Collection)**

A car heads southwesterly towards Masontown and Martin from Brownsville Junction June 1, 1947. The railroad under the trestle was the Dunlap Creek Branch of the Monongahela Railroad, and the trol-ley track leaving to the left of the photo went to Brownsville. **(James P. Shuman)**

BROWNSVILLE LINE

Cars operated over joint trackage with the Martin Line from Uniontown to Brownsville Junction, 7.1 miles. Turning right at the junction, the cars offered hourly service to Brownsville, a total distance of 16.6 miles. The route was notable for some spectacular bridges. The line was abandoned January 28, 1950, a week before the Martin line.

Car 720 operates southbound (towards Uniontown) into Davidson siding on the Brownsville line, just north of the Allison Works. Allison was one of the area's largest coke manufacturers. Summer has already arrived on May 29, 1949, but the Brownsville line would not survive another winter.
(James P. Shuman)

The last service on the Brownsville line occurred on January 25, 1950. A week later, on February 4, service was terminated on the Martin line as well. A special car, number 720, was chartered to tour the Martin lines on February 5th because the Fairchance line would be abandoned by the end of March. The special is stopped for a photo op west of Uniontown en route to Brownsville Junction and Martin.
(Eugene Van Dusen Collection)

Two small towns between Brownsville Junction and Martin on the Martin line were photogenic in their own right principally because of the precarious grade crossings and hilly terrain in the area. Motorists encountered West Penn interurbans at unsignalled crossings when they least expected them and the trolley motormen had to exercise the utmost of care when negotiating the crossings. This was further exacerbated by the absence of air and air whistles or chimes on the cars. The gong was the only means of warning their approach. Even regular riders peered over their shoulders out of the car windows when crossing the spidery trestles in the area.

Car 720 is southbound on Cats Run trestle, about a mile south of Masontown on February 5, 1950. Pennsylvania Route 166 passes under the trestle. The Monongahela Railway's Cats Run branch went under the trestle and crossed the highway at grade just to the right of the blue auto in the picture. The trestle has a descending grade of 5.5% in the southbound direction, which is fairly steep.
(James P. Shuman)

This will be the last time a West Penn car will endanger motorists in McClellandtown, since the 720 is operating a farewell fantrip on the day following formal abandonment of the Martin line, February 5, 1950.
(Eugene Van Dusen Collection)

CITY CARS

In the 1940s, West Penn operated several city cars numbered in the 280, 290, and 610-series on local lines in Connellsville, in addition to operating these cars on inter-city routes such as from Greensburg to Trafford. The 600 series deck roof cars were withdrawn from service in August, 1939 when the 831-842-series Cincinnati lightweights were brought down from the Allegheny Valley division and pressed into service on the South Connellsville line. Four of the 600-series cars, 610-613 were retained in storage until the war's end in 1946, just in case they were needed.

Car 611 was brought out of storage on August 31, 1941 for this NRHS convention special, shown here making a photo stop at the Connellsville station. The car was a veteran of the South Connellsville line prior to its retirement. The NRHS convention was held in nearby Pittsburgh that year. The 611 was originally Westmoreland Railway 2, built by St. Louis Car Company in 1905. It was later West Penn 242, then 611. (**James P. Shuman**)

A passing siding close to the Pennsy main line on Clay Avenue extension in West Jeannette allows 286 to meet 292 on July 5, 1941. Even though these cars are signed up for Irwin, some cars operated through from Greensburg to Trafford until August 15, 1942. (**James P. Shuman**)

CAR 739, THE "FAYWEST"

West Penn 739, the last 700-series car built new in the Connellsville Shops (1925) was converted two years later, in 1927, to a center-entrance, parlor car. The 700s were all originally built as two man cars, but most of this series (all but 11) were converted to one-man operation prior to 1933 by adding a front entrance. One car (700) was scrapped prior to the conversion program. Car 739, fittingly, was the last car to operate, on August 10, 1952, and was the last carbody to be rescued by the Pennsylvania Trolley Museum near Washington, PA.

Late in the afternoon of August 10, 1952, the last West Penn 700-series car to be built at the Connellsville shop makes a last stop on the Uniontown station loop to discharge the last passengers. The end has come. However, on a happier note, this car may yet run again at the Pennsylvania Trolley Museum, albeit on a set of former Pittsburgh Railways Brill trucks. The car's original name "Faywest" was a contraction of the two county names, Fayette and Westmoreland. **(Edward S. Miller)**

CAR 832 — A SURVIVOR

The only passenger rail car saved on its own trucks and operated under its own power to the Pennsylvania Trolley Museum, in 1953, was West Penn 832. In addition to being one of the last Cincinnati curved-side cars to have been built, it was the only example of this type car to be saved by any museum, intact. Ironically, other museums around the country have managed to resurrect carbodies used as shed and homes to all recreation of cars of this design. The West Penn car was in badly deteriorated condition when restoration began, such that it has never been operated at the museum where it is preserved.

On May 24, 1949 the 832 was between assignments at the Connellsville Shop when the late Charles E. England, then a member of the Pittsburgh Electric Railway Club and one of the founding fathers of the museum effort, snapped this photo of it. **(Charles E. England, courtesy of Pennsylvania Trolley Museum)**

Johnstown Traction Company
Johnstown, Pennsylvania

Johnstown, PA in 1940 was a city of approximately 75,000 population located along the Pennsylvania Railroad main line about 65 miles east of Pittsburgh. It was roughly midway between Altoona and Pittsburgh, and like Pittsburgh its main industrial base was production of steel. This type of city was a natural for public transit usage and traction was still king there. As of 1940, most of the Johnstown Traction Company trolley routes were still intact. Only the interurban line to Windber and Somerset had been abandoned, it having been a victim of the 1936 flood. The Southern Cambria interurban route east to Ebensburg (a separate operation) had also been previously abandoned. During the decade of the 1950s, JTC operated 65 cars over 30 miles of track.

Lines remaining in 1940 were Morrellville-Oakhurst, Coopersdale, Franklin, Moxham via Horner Street, Ferndale-Benscreek, Roxbury, and Southmont.

The Horner Street line was converted to trackless trolley on November 20, 1951, and the Southmont line to diesels about 1954. A replacement diesel bus line to Dale began in August 1940. Windber buses operated over an entirely different route than the trolleys had, via Benscreek.

The remaining Benscreek line consisted of a rush-hour only operation in the mid-1950s, and bad track forced its abandonment without bus substitution, in 1957. The Oakhurst line was a one-car shuttle extending from the outer end of the Morrellville line to Oakhurst, and became a mere franchise run in the 1950s. The track was still available for fantrips as late as 1958. In 1955 the heaviest lines operated on 15-20 minute headways with half-hourly service on Sundays and Holidays. By the 1958 recession in the steel industry, headways had shrunk drastically.

In 1958 several trolley coaches were purchased from Wilmington, Delaware, and Covington, Kentucky. They were gradually refurbished and the entire streetcar system was abandoned on June 11, 1960. The trolley coaches continued to run until they too were disposed of, on November 11, 1967. The 1951-model St. Louis Car Company trolley coaches were sold to Mexico City and at least two of the ex-Delaware Coach Company pre-war vintage Brills were preserved in museums.

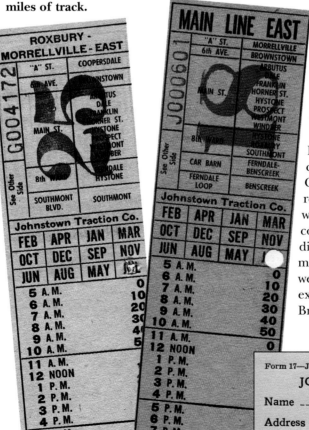

Form 17—J. T. Co.

JOHNSTOWN TRACTION COMPANY

Name _____

Address _____

Date_____ Time_____

Conductor No._____ Car_____

Amount dropped in_____Amount to be refunded_____

Signed by Conductor_____

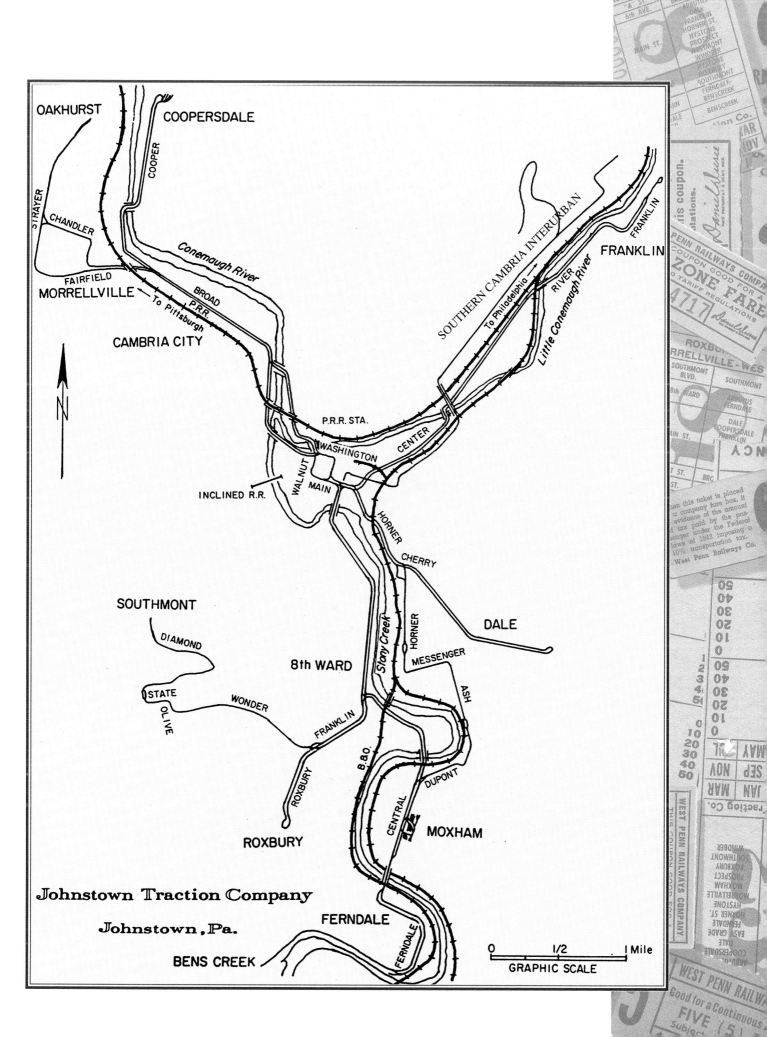

OAKHURST COOPERSDALE

STRAYER
CHANDLER
FAIRFIELD
MORRELLVILLE
To Pittsburgh
Conemaugh River
BROAD
P.R.R.
CAMBRIA CITY

N

SOUTHERN CAMBRIA INTERURBAN
To Philadelphia
Little Conemaugh River
RIVER
FRANKLIN
FRANKLIN

P.R.R. STA.
WASHINGTON
CENTER

INCLINED R.R.
WALNUT
MAIN

HORNER
CHERRY

SOUTHMONT
DIAMOND

STATE
OLIVE
WONDER
FRANKLIN

8th WARD

Stony Creek
HORNER
MESSENGER

DALE

ASH

ROXBURY

B. & O.
CENTRAL
DUPONT
MOXHAM

ROXBURY

Johnstown Traction Company

Johnstown, Pa.

FERNDALE
FERNDALE

BENS CREEK

0   1/2   1 Mile
GRAPHIC SCALE

MOXHAM-FERNDALE LINE

Summer in Johnstown. Just outbound from Moxham Barn, the double tracked Ferndale/Benscreek line necked down to one track to negotiate the Stony Creek Bridge. Car 311 is seen at Central Avenue and Jacob Streets on June 30, 1957, about to enter the single-track private right-of-way.
(Edward S. Miller)

Fall in Johnstown. PCC car 416 ambles toward Ferndale loop on a brick-lined trackway. There were few passengers enjoying the fall foliage on October 12, 1958, because many of the residents were standing in the unemployment lines during that recession wracked year.
(William D. Volkmer)

Winter in Johnstown. On December 6, 1958, the 405 is coming inbound approaching Moxham. At this point the single track over the Stony Creek Bridge becomes double track along Central Avenue in Moxham.
(William D. Volkmer)

On a snowy January 17, 1959, PCC 404 turns off the private right-of-way of the Ferndale line and heads up the center of the road towards Ferndale loop. The PCCs performed well in the snowy climate of Johnstown. **(W.D. Volkmer)**

Later that same day, car 407 is caught inbound off the Stony Creek bridge about to enter Central Avenue, Moxham and a return to double track operation. **(W.D. Volkmer)**

Ferndale Cars were alternately through routed to Morrellville and Coopersdale on their return trips. At Main and Market Streets in downtown Johnstown, PCC 416 accelerates towards Ferndale on Friday, August 27, 1954. The familiar Pepsi Cola bottle cap signs had yet to adorn the cars as they did in the later years. **(Edward S. Miller)**

Even though car 414 is still outbound and a mile or so shy of the Ferndale loop, the Operator has already cranked up the destination roller to read Coopersdale, lest he forget where his next trip destination is. The date is Friday, August 27, 1954. **(Edward S. Miller)**

The truss bridge over Stony Creek was an ideal spot to shoot trolley cars in Johnstown because of its east-west orientation and scenic setting.

Former Bangor, Maine double truck Birney 311 is outbound over the bridge on June 30, 1957. (**Edward S. Miller**)

PCC 405 crosses on July 27, 1958 en route to Ferndale. (**W.D. Volkmer**)

FRANKLIN LINE

The Franklin line was one of Johnstown's first transportation arteries, having been opened as a horsecar line in 1883, and of course, was wiped out in the great flood of 1889 along with the entire rest of the city. It was rebuilt and extended to Franklin in 1896, temporarily converted to bus in 1958 for bridge construction, and converted to trolley coach in 1960. Fifteen-minute service on the route was augmented with tripper cars during shift change times at the steel mills. An extension of the trolley coach line was made into neighboring Conemaugh in 1962 as the route served several Bethlehem Steel mills.

Double truck Birney car 311 has crossed the Conemaugh and Black Lick railroad tracks on Clinton Street and traverses the Conemaugh River bridge prior to entering Maple Avenue en route to Franklin, on June 30, 1957. (**Edward S. Miller**)

There was a short stretch of single track near the outer end of the Franklin line that was unprotected by any sort of signal system. PCC 404 approaches the Franklin loop on May 23, 1959 having successfully navigated that "dark" territory without a meet with an inbound car. (**W.D. Volkmer**)

The Franklin line was converted temporarily to bus operation during the summer of 1958 and into the spring of 1959 to allow for construction of a new bridge on Maple Avenue spanning the railroad entrance to the Bethlehem Steel Company. Cars could operate as far as the city side of the bridge and would turn back on a temporary crossover at that point. Car 350 is doing just that on July 12, 1958. **(W.D. Volkmer)**

Car 350 and 356 operate along Clinton Street near the B&O Railroad crossing, inbound on the Franklin line during the NRHS National Convention, September 6, 1959. The B&O branch line ended right here at this mill, but the conventioneers who had traveled to Johnstown via Somerset on the B&O had detrained at the Ferndale crossing to board the waiting streetcars. **(W.D. Volkmer)**

MAPLE AVENUE BRIDGE

*Here are car 352 and others crossing the newly completed highway bridge to Franklin. The multiple car crossing was occasioned by the final day of operations festivities. (**W.D. Volkmer**)*

PCC car 404 crosses the newly completed Maple Avenue Bridge on May 23, 1959. During the bridge's construction there was much conjecture (and probably debate) as to whether or not trolley tracks would be installed on the bridge given the imminent demise of the trolley system. To virtually everyone's amazement, the tracks were installed and the streetcars traversed the bridge for a little over a year before trolley coaches were substituted.
(W.D. Volkmer)

ROXBURY LINE

The Roxbury line was originally built to serve Luna Park then located at the end of the line. It appeared in the latter years to be Johnstown's strongest line as it enjoyed the most frequent service of any of the routes. Its signature trackage was a short stretch of center-of-the- (road?) private right-of-way up Roxbury Avenue that was finally paved for trolley coach operation in 1960.

Special Civil Defense advertising car 401 approached the Roxbury loop on Roxbury Avenue on July 12, 1958.
(W. D. Volkmer)

The 311 rumbles across the Franklin Street bridge as it and a PCC following exit the downtown area of Johnstown on June 30, 1957. (Edward S. Miller)

A few months later, October 12, 1958, the fall foliage is in full swing as the 415 passes the same location.

SOUTHMONT LINE

The Southmont line seems to have been the least photographed in color of all the Johnstown routes. It was one of the most scenic and was always operated with double-end equipment in contrast to most of the other lines that had loops at the end. Residents of this area of town had the option of taking a shorter route to town via the Westmont inclined plane and transferring to a trolley in the downtown area.

Car 211 is inbound along the side of Olive Street at State Street on October 9, 1953. This series car would be phased out on abandonment of the line. **(Eugene Van Dusen Collection)**

Car 214 seems to be engulfed in the brilliant fall foliage as it wends its way towards downtown alongside Southmont Boulevard, also on October 9, 1953. **(Eugene Van Dusen Collection)**

The Southmont line had long been abandoned on July 19, 1959 when this fantrip car backed up onto the line as far as the section break in the overhead would permit in order to allow this "fantasy shot". PCCs such as the 407 never operated on the line because there was never a loop constructed at the outer end. (**W.D. Volkmer**)

Car 311 is preparing a similar caper, about a year earlier, on July 27, 1958. Sporadic street repairs in the area were beginning to cause the flangeways to fill up with asphalt in spots at this time, but the venturous fans bumped over the pavement with the erstwhile Birney to get this photo! (**W.D. Volkmer**)

SOUTHMONT JUNCTION

The 413 and 498 pass en route to and from Roxbury at Southmont Junction on the last day, June 11, 1960. Trackage was still powered a couple hundred feet up the Southmont line from this point in order to get disabled cars off the line and out of the way until they could be either repaired or towed back to Moxham barn. (**W.D. Volkmer**)

MORRELLVILLE LINE

Between 1911 and 1915, the Morrellville line was isolated from the rest of the JTC system by the PRR tracks. Passengers were required to get off the Coopersdale cars and walk across the four track main to catch a ride up into Morrellville and Oakhurst. An underpass under the Pennsy on Fairfield Aveenue constructed in 1915 allowed for the JTC cars to operate into that borough. The resulting Morrellville line consisted of one enormous unidirectional loop that toured Chandler, Strayer and Fairfield Streets.

Flanked by vintage 1960-era automobiles, car 417, the highest numbered PCC, now stripped of its Pepsi sign, rolls outbound on Roosevelt Boulevard. The car will soon turn left onto Broad Street for one last trip to Morrellville on June 11, 1960, the last day of service.
(Robert E. Bruneau, Bill Volkmer Collection)

Here PCC 402 breezes inbound up Strayer Avenue on August 15, 1956. In the background the Oakhurst bus can be seen making an outbound run.
(Raymond E. Mc Murdo, Bill Volkmer Collection)

On June 11, 1960, the last day of trolley operation, everything that could turn a wheel in town operated. Company rules that forbade two cars from climbing the hill in Morrellville together (in order to keep the power demand charge low) was waived for the occasion. Thus, the 352 and 355 made it to Morrellville together, allowing this photo.
(W.D. Volkmer)

Car 356 negotiates the strange S curve at the intersection of Decker and Strayer Street during a fantrip on July 27, 1958. Regular service over this feeder line was discontinued in 1953, but the track was kept in active condition, for use on fantrips, almost until the end of service in 1960. Sometimes the railfans had to shovel dirt out of the flange-ways in order to make the run possible though! **(Charles Ballard.)**

The author's 1958 Plymouth was "posed" along-side the 350 making its last run prior to being shipped off to Arden Museum (since renamed the Pennsylvania Trolley Museum). The location, Corrine Avenue on the Oakhurst Line, was recorded on film May 23, 1959. **(W.D. Volkmer)**

PENNSYLVANIA

The Oakhurst Shuttle was a short single-track segment that began at the outer end of the Morrellville line and operated totally without the convenience of a single passing track. Outbound it ran on Strayer Street and Corrine Avenue, terminating at Daniel Street. On July 27, 1958 car 356 changed ends at the end of the line. **(Charles Ballard)**

COOPERSDALE LINE

Early on the morning of February 28, 1959 PCC 413 has few customers as it prepares to make a right turn over the Conemaugh and Black Lick Railroad tracks, headed inbound towards downtown Johnstown. Trolley service would be an "on again, off again" proposition until finally abandoned in the fall of that year. **(W.D. Volkmer)**

Ex-Bangor car 311, one of 14 cars purchased during World War II from Bangor, Maine and New Haven, Connecticut is seen inbound crossing the C&BL RR tracks at grade near the junction with the Morrellville line on July 12, 1958. **(W.D. Volkmer)**

The loop at Coopersdale was the site of the long-abandoned Coopersdale car house, where unused cars were kept in dead storage as a parts supply for the remaining non-PCC cars in the fleet. Car 408 lays over at the loop on August 16,1958, a Sunday afternoon, when two cars were adequate to cover the four JTC routes then in operation. **(W.D. Volkmer)**

At the same Coopersdale Loop location on April 4, 1965, the Pennsylvania Trolley Museum-chartered trolley coach 710 illustrates "before and after" scenario to the above photograph. The fans realized that even the trolley coach days were numbered in Johnstown and wished to record the action for posterity before it became too late. **(W.D. Volkmer)**

Downtown Johnstown

The principal thoroughfares in downtown Johnstown were Main Street and Washington Street. Because of a one-way street program, the trolleys made a loop in the downtown area with a few block stretch of bi-directional traffic on Main Street. Car 407 was snapped from the front vestibule of an opposing car on Main Street on May 23, 1959. *(W.D. Volkmer)*

Cars 402 and 415 pass City Hall at Main and Market Streets on Saturday August 28, 1954. (Edward S. Miller)

Cars exited downtown Johnstown southward on Franklin Street destined for Roxbury, Ferndale, Southmont and Moxham. Here we see the 311 outbound on Franklin Street at Vine on June 30, 1957. **(Edward S. Miller)**

Turning from Washington Street into Walnut, the 214 passes a classic stainless steel diner of the 1940s era. The date is June 30, 1957, and gas is still a quarter a gallon! **(Edward S. Miller)**

WESTMONT BOROUGH INCLINE

Johnstown's "other" traction system, the Westmont inclined plane can be seen on the hill-side climbing the opposite side of Stony Creek. A nice, sunny Sunday, June 30, 1957 allowed this close-up shot of the action on this inclined plane. This outdoor elevator is still in operation as this is written. *(Edward S. Miller)*

The 311 is about to turn from Walnut Street onto Main on April 23, 1960. The inclined plane took autos and city buses up to the Southmont area (for a fee) obviating a lengthy and circuitous course by road. *(W.D. Volkmer)*

On Saturday August 28, 1954, car 355 operated outbound at Riverside passing siding (obviously long out of service).
(Edward S. Miller)

This is as far down the Benscreek line the 311 could operate on May 23, 1959, some two years after the line had been abandoned. Much of the rail had not yet been taken up however.
(W.D. Volkmer)

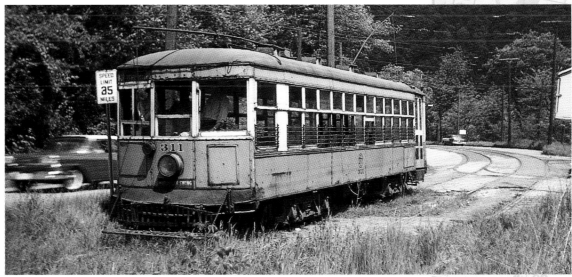

In the days before mega-gas stations, Bill's Gas Station serves as the backdrop in 1953 while the 356 changes ends at the Benscreek terminal. Cars formerly continued on to Windber from this point but the 1936 flood ended that operation forever.
(Volkmer Collection)

The B&O Crossing on the Ferndale Line

The cause for the premature abandonment of the Ferndale line was the condemnation by the B&O railroad of the crossing diamond just south of the river bridge in Moxham. Because the B&O unceremoniously removed the diamond in November 1959, there was no formal last day operation of the line. Abandonment of the Coopersdale line was carried out at the same time.

The Ferndale line operated past Moxham carhouse and crossed over Stony Creek and the B&O Railroad on a short stretch of private right-of-way. The 405 makes the obligatory safety stop at the B&O before proceeding across on July 27, 1959. (W.D. Volkmer)

Car 311 approaches the B&O crossing on the Ferndale line and an outbound PCC car. The date is May 23, 1959 and unbeknownst to the JTC at the time, the crossing diamond ahead would be the cause of the premature abandonment of this line in a few months. (W.D. Volkmer)

On March 14, 1959, the author managed to catch the tail end of a B&O freight at the crossing with PCC 408 beyond the diamond, waiting for it to pass. (W. D. Volkmer)

Moxham Barn serves as the mustering point for cars 403, 404 and 417 on July 17, 1959. Preceding the day's fantrip, the author has changed the PCC destination signs to two places that the PCCs never ran, and one (Horner Street) where there had been no cars for over eight years. (W. D. Volkmer)

During the period of 1958 to 1960, when abandonment finally occurred in Johnstown, it seemed that the principal source of revenue for the system was from railfan trips. The price for a trip was so low the fans couldn't afford NOT to take advantage of the offer. In the summer of 1958 when the author first moved to Altoona, there were only two cars in regular service on Sunday, operating on 1½-hour headways over the two lines that were in service, Roxbury-Ferndale and Morrellville-Ferndale. The nation was in a deep recession at the time, and the Coopersdale line was bused on the weekend and the Franklin line was truncated because of the reconstruction of the Maple Street Bridge and hence it too was operated entirely by bus. Because I lived in Altoona, about an hour's drive over the mountain, I seldom passed up the chance to take in what had become about a twice-a-month event, a trolley fantrip over America's last remaining small town streetcar operation. The motorman was almost always a gentleman by the name of Johnny Yuchek, who was sympathetic to rail fans. As such, he would always let the fans motor the cars as long as they were in the outlying districts of the city. To my knowledge, I and one other fan were the only people Johnny allowed to motor the cars in the downtown streets, owing to my frequent ventures on the line. On the following pages, the reader will be treated to many photos that were made on those fantrips, preserving the system for posterity via color film.

On the last day of streetcar operation, June 11, 1960, a small group of fans including the author chartered trolley coach 731, the very last electric vehicle ever constructed by Brill, and went for a ride in parallel with the streetcar fantrips of the day. Accordingly, the author installed a Dayton trolley coach destination roller sign, just to make the pictures more interesting! At Main and Market Streets, PCC 408 makes one of its final runs past the 731, signed up for Dayton's "Ohio Street." (W. D. Volkmer)

Moxham Via Horner Street Trolley Coach Line

St. Louis-built trolley coach 701 shared the same stop with PCC car 405 on Washington Street, at the corner of Walnut Street downtown. The 405 was a fantrip car mythically signed up to Benscreek. Since Benscreek was always without a loop at the end, no PCC car ever operated there, but for some unexplained reason, the destination was included on the PCC roll signs. Possibly, a loop had been planned at one time. The date of the fantrip was July 27, 1958. *(W.D. Volkmer)*

One of the benefits of this charter was an opportunity to travel the Horner Street line. Here the fantrip poles have been pulled in order to allow St. Louis TC 705 to pass inbound along Ash Street. *(W. D. Volkmer)*

Old St. Louis car 214 and a Horner Street trolley coach are seen at the Washington Street location, but from the rear, on June 30, 1957. A Pittsburgh fan had chartered the 214 on this date. *(Edward S. Miller)*

THE CAR ROSTER

211-220 ST. LOUIS CAR COMPANY 1916 SCRAPPED 1945-57

Double end car 219 turns at the Franklin loop in the 1940s. The loop was situated at Main and Bon Air Streets in Franklin.
(Frank Watson)

Car 214 operates inbound on Broad Street near Fairfield Avenue on Sunday June 30, 1957. All cars of this series were stored awaiting scrapping the next year following the demise of the Benscreek shuttle car operation.
(Edward S. Miller)

On the rip track at Coopersdale Carhouse, car 211 awaits its fate on July 19, 1959. Its working days are over.
(W.D. Volkmer)

221-230 St. Louis Car Company 1917 scrapped by 1959

Car 223 is stored at Moxham Barn on August 27, 1954. Its useful days were numbered at this point in time. The principal use of the double-enders, other than fantrips of course, was for the Benscreek shuttle line. (**Edward S. Miller**)

305-311 Ex-Bangor, Maine Wason Car Company 1920

Car 311 represented the last product of the Wason Car Company of Springfield, MA to operate on any transit system in the United States. The car is shown near the Roxbury Loop, on July 27, 1958. Following abandonment of the system, the car was sent to Rockhill Trolley Museum where it is preserved to this day. (**W D. Volkmer**)

350-369 St. Louis Car Company 1926 Built as single end.
350-358 Converted to double end.

Car 364 is passing Moxham carhouse, circa 1945. (*Frank Watson*)

Following cessation of streetcar service in Johnstown, many of the remaining cars in this group were preserved in museums. Car 350 went to the Pennsylvania Trolley Museum. The 351, is reported to have been saved in California by a private individual and may someday operate on Market Street, San Francisco. Car 352 is at the National Capital Transportation Museum, Wheaton, Maryland. The 355 went to Rockhill Trolley Museum, Orbisonia, PA. The 356 and 357 are at the Shoreline Museum, Branford, Connecticut. The 358, converted to a diesel rail car by its former owner, the Stone Mountain Scenic Railway, in Georgia, is currently being restored at the Trolley Museum of New York near Kingston. Car 362 is reportedly at the Fox River Museum near Carpentersville, Illinois.

The 355 operates on what was referred to as the "Main Line" by Johnstown Traction, on Franklin Street, between downtown and the Valley Street Junction between the Roxbury and Ferndale lines. The junction was known in local parlance as the "Eighth Ward Transfer". This shot was made on the author's "personal fantrip", April 23, 1960, a month and a half prior to the line's closure. (*W.D. Volkmer*)

PENNSYLVANIA TROLLEYS
ONE FARE
Johnstown Traction Company

056398

401-417 PCC Cars — St. Louis Car Company, 1947

Car 401 had a special paint job for the Civil Defense initiative during the cold war. By 1958, when this photo was made at Moxham Carhouse, she was in need of another paint job. **(W.D. Volkmer)**

PCC 405 lays over at the Ferndale Loop on a frigid January 17, 1959. A recent television program dealing with collectibles indicated that 1950s era Pepsi Cola signs are highly sought after by collectors. They all disappeared off the trolleys in Johnstown just before the last run so the collectors must have had an insight into the business. **(Charles Ballard)**

The sign on the rear of car 408 read "Ride This Car - Keep it Running." Apparently this worked to some degree, but inevitably, the car lines had to be replaced by trolley coaches. This shot was taken at Roxbury loop on August 16, 1958. **(W. D. Volkmer)**

THE TROLLEY COACH IN JOHNSTOWN

On August 27, 1964, all streetcar rails had been removed and St. Louis coach 702 rounds the bend on Roosevelt Boulevard en route to downtown from Morrellville. (**W. D. Volkmer**)

The story of traction in Johnstown would not be complete without a section devoted to the trolley coach era in Johnstown. In addition to being the last city to adopt the PCC car as the vehicle of choice, Johnstown entered the trolley coach era in 1951, almost at the close of *that* transportation fad. Even more astonishing was the fact that in 1958, an entire fleet of second-hand coaches was purchased from Wilmington, Delaware (Delaware Coach Company) and Covington, Kentucky (Cincinnati, Newport and Covington Street Railway) in order to replace the streetcars! In the pages that follow, we shall briefly visit the trolley coach action in Johnstown between November 20, 1951 and November 11, 1967, the last day of service for this mode.

A short while earlier, the 702 had passed Brill 710 on Broadway. The 710, formerly of Delaware Coach Company, was working the Coopersdale line. (**W. D. Volkmer**)

Brill 710 rests at the Franklin loop on April 4, 1965. Only four of the little Brills were purchased and put into service in Johnstown. This particular coach is now preserved at Rockhill Trolley Museum, in central Pennsylvania. **(W. D. Volkmer)**

On the last day of revenue service, November 11, 1967, the famous driver (of streetcar fantrip fame), Johnny Yuchek turns the corner of Washington and Walnut Streets aboard St. Louis coach 704. None of these coaches were preserved and they were the last model produced by the builder. **(W. D. Volkmer)**

On November 14, 1959, newly refurbished Marmon-Herrington trolley coach 740, formerly Cincinnati, Newport and Covington 653, must wait another seven months before being pressed into revenue service. Coach 735, a Brill, also from Covington (ex- 663) sits adjacent to it in the Moxham storage yard. **(W.D. Volkmer)**

Brill 732 makes good use of the paving on Roxbury Avenue where the trolley rails once stood, as she approaches Roxbury loop on April 29, 1961. The old Pepsi Cola signs of the streetcar days have given way to 7-Up! **(W.D. Volkmer)**

After the streetcar system was abandoned, JTC realized that they had enough trolley coaches to convert the Conemaugh bus route to trolley coach. By doing this, coaches on the Franklin line could alternate between the Conemaugh loop and the Franklin loop and give the same level of service to the steel mill. On the last day of service, November 11, 1967, coach 705 lays over at the end of the Conemaugh line. *(W. D. Volkmer)*

A parade of last-run trolley coaches on Walnut Street, at Main, is led by JTC 741, ex-CN&C 655. Behind it is a St. Louie, another Marmon, and a GM diesel bus. **(Ken Douglas, Bill Volkmer Collection)**

On the last day of service, a Saturday, many of the trolley coaches were already declared surplus, and lined up in "Brill heaven" at the power plant on Horner Street. In a few hours the rotary converters would be ceremoniously silenced forever, ending 75 years of electric traction history in Johnstown. **(W.D. Volkmer)**

MOXHAM CAR HOUSE AND SHOPS

Car 405 prepares to depart the carhouse as the 2nd car on the fantrip of July 27, 1958. **(Charles Ballard)**

PCC 410 peeks out from behind the paint shop and the carbarn on Saturday June 30, 1954. **(Edward S. Miller)**

The late Stephen D. Maguire made this photo of car 403 and an ancient Yellow Coach at Moxham Barn on August 31, 1955. **(S.D. Maguire, Volkmer Collection)**

FINAL DAY CEREMONIES JUNE 11, 1960

It was a warm summer's day when car 412 was decorated for the last run at Moxham Carhouse. This car was probably chosen to receive bunting covering its front end because of the cancerous rust engulfing the car, making the car otherwise largely unphotogenic! **(W. D. Volkmer)**

A large crowd of dignitaries has gathered in front of City Hall at 5 PM Saturday afternoon June 11, 1960 culminating a day of fantrip and regular cars with railfans and citizens alike taking one last ride on the trolleys in Johnstown. Bunting draped PCC 404 leads the parade with Birney 311 bringing up the rear, to be the very last car to operate in Johnstown. **(W. D. Volkmer)**

Pittsburgh Railways Company
Pre-1964
Port Authority Transit
Post-1964
Pittsburgh, Pennsylvania

The Pittsburgh Railways Company could trace its roots back to 1859 when a single horsecar route began to operate along Penn Avenue. Trackage, in common with many Pennsylvania horse and trolley lines, was laid to 5 foot $2^1/_2$ inches. The hilly terrain of metropolitan Pittsburgh set the stage for some rather spectacular railway construction. Steep grades (Rt. 21-Fineview, for example, had a 12.16% incline on Henderson Street between Federal and Boyle), spidery trestles on several routes, long suspension bridges over the city's three rivers, and inclined planes carrying both passengers and motor vehicles added color and uniqueness to the system.

In 1940, nearly all of the rail lines were still operational, in an age where many cities had either partially or totally converted to rubber tired vehicles. The Railways was an early convert to PCC cars, and when World War II broke out, 301 PCCs were already on the streets with 100 more on order. Fortunately because of the importance of the area's industries to the war effort, the government allowed cities such as Pittsburgh, Boston, and Philadelphia to continue to amass large fleets of new streamliners. Hence, 65 more were ordered in June 1942 and 100 more in January 1944. Counting the 200 cars received after the war's end, Pittsburgh Railways operated a total of 666 PCC cars, more than any other city in America except Chicago.

Non-PCC car operation in Pittsburgh during the 1940s and early 50s was limited to a few hundred "low-floor" cars, of a common deck-roof design dating to 1910, which were numbered largely in the 4200 to 5500-series.

As riding slacked off in the early 1950s due to increased automobile usage, the Railways Company began pruning the roster and non-profitable lines such as the interurbans to Washington, Charleroi/Roscoe, during the summer of 1953. The small shuttle lines around town were also discontinued at this time such as the Evans Avenue, Corey Avenue,

Thornburg, Bon Air, Reedsdale, Schoenville, Atwood, Oakmont, Homeville, Evergreen, and P&LE Transfer. Lines to Etna and Millvale were also abandoned in 1952.

By 1956, only the PCC cars were left in service over 300-plus miles of track. Beginning in the summer of 1957 with the abandonment of several northside lines, the Railways began a slow, but steady conversion process to diesel buses. Individual line conversion was often driven by bridge replacements without car tracks, such as the Point Bridge in June 1959, the Sharpsburg (62nd Street) Bridge in November 1960, and others.

Although Pittsburgh Railways Company was not bankrupt, the management was not inclined to spend additional monies on rail cars or buses. They could foresee the inevitable public ownership on the horizon. The Port Authority of Allegheny County, a public agency, took control on March 1, 1964 to own and continue to operate the system. Car line conversions to diesel bus were completed over most of the remaining system to the point where by November 13, 1971; the only remaining rail routes were the three South Hills lines. The Library, Drake, and Dormont were kept rail, plus a new route, 49-Arlington-Warrington, over the top of Mt. Washington to preserve trackage as a diversion for trolley tunnel traffic, when the tunnel was out of commission for repairs.

In the 1980s, Port Authority Transit, as it became known, upgraded the Dormont line to light rail standards, dug a subway under downtown Pittsburgh, purchased new articulated railcars, and extended the line to a new loop and shop complex area, at South Hills Village. The new line, fully opened in 1987, incorporated the old Route 38A trackage from Mt. Lebanon to Castle Shannon. In the late 1990s, work has begun on rebuilding the long out-of-service interurban trackage between South Hills Junction and Castle Shannon via Overbrook to light rail standards.

WASHINGTON INTERURBAN LINE

Cars assigned to the interurban line from Pittsburgh to Washington, PA covered 29 miles in 1 hour and 49 minutes. Prior to 1949, when PCCs were purchased, there were three varieties of cars plying the route. They were: 1) Railroad roof Brill center-exit cars numbered 3700-3714, 2) single ended deck-roof cars 3750-3769 built by Standard Steel Car Company in 1925, and 3) the speedy arch-roof St. Louis Car Company cars built in 1928, and numbered 3800-3814. The 1949 PCC cars 1700-1724, plus a few modified 1600-series cars, 1613-1619, and 1645-1648, that were operated over the Washington interurban line, until the lines were truncated by two-thirds. The shortened Charleroi line, to Library, began June 28, 1953, and the shortened Washington line, to Drake, began August 29, 1953, ending the interurban era in Pittsburgh, for all intents and purposes.

Center-exit car 3704, a J.G. Brill Company product of 1916 is southbound on the outskirts of Washington, en route to Pittsburgh in June 1944. These cars were replaced by interurban PCC cars five years hence. (**Eugene Van Dusen Collection**)

Elco loop in Roscoe, PA represented just about as far as one could get from downtown Pittsburgh by PRC interurban car. Red interurban car 3802 makes the turn on July 3, 1948, about five years prior to abandonment of the route beyond Library. (**Eugene Van Dusen Collection**)

Car 3801 passes Washington Junction en route to Charleroi in June, 1944, a month celebrated for the landing on Normandy Beach during World War II. **(Eugene Van Dusen Collection)**

Inbound from Washington at Washington Junction is car 3704, one of the World War I era interurban cars, in June 1944. **(Eugene Van Dusen Collection)**

WASHINGTON, PA CITY OPERATIONS

Washington, a small city some 29 miles south of Pittsburgh was essentially second runner-up to Johnstown as one of America's last small city trolley operations. Until the early 1950s there were two cross-town car lines, the East and West (2.73 miles), and the Jefferson-Maiden (3.47 miles) plus a one-mile route to North Washington. Until 1953, the city was linked to Pittsburgh via the Washington interurban line.

Car 4355 has changed ends at the east end of the East and West line on E. Beau Street in front of East Washington High School, just east of Morgan Avenue on May 20, 1950. The East & West line was abandoned on May 16, 1953, about three months prior to the discontinuance of the interurban line. **(Edward S. Miller)**

In North Washington, double end car 4383 rests at the end of the line, on October 10, 1952. The Washington local lines would be abandoned the following year. **(Gordon E. Lloyd, Eugene Van Dusen Collection)**

Car 4361 operates on the East and West line in downtown Washington, May 20, 1950.
(Eugene Van Dusen Collection)

Car 3750 rests in the Tunnel Car House Yard over the July 4th holiday in 1948. A steel belt rail seals the left-hand door, which dated from the car's service on Rt. 23-Sewickely, where single track at roadside made use of the right-hand door unsafe when running with doors to road.
(Eugene Van Dusen Collection)

DOWNTOWN PITTSBURGH

Here are some scenes of downtown Pittsburgh to get an idea of what the city looked like in the 1950s before skyscrapers and subways became the norm.

Car 1724 drops off passengers from Library in front of the William Penn Hotel on Grant Street at 6th Avenue. The date was September 10, 1956 and the airporter buses were almost exclusively the bullet ended Flexibles such as the one on the right in the photo. **(W.D. Volkmer)**

Interurban car 1716 turns onto Liberty Avenue from Grant Street in the middle of the city loop for the Library cars, on August 24, 1954, a little over a year after the Charleroi line was truncated to Library. In addition to the bar pilot and roof headlight, the interurban PCCs in the 1700-series were equipped with Clark B-3 trucks that were better suited to private right-of-way operation.
(Edward S. Miller)

A Rt. 44-Knoxville car, the 1621 stops adjacent to Penn Station in downtown Pittsburgh on June 29, 1958. The umbrella shed would soon be removed as a part of a general reconstruction of the area that resulted in the Greyhound station being built across the street. **(W.D. Volkmer)**

Pittsburgh's third PCC car, the 1001 holds down the Rt. 88-Frankstown outbound at Liberty Avenue and Grant Streets on June 14, 1959. This route, which at one time had MU low-floor cars, operated inbound on Penn Avenue and outbound on Liberty. **(W.D. Volkmer)**

Closer to the Point, car 1537 has just come onto Liberty off 6th Street on July 17, 1960. The destination sign Trafford Express is more or less bogus, as this was a fantrip using a fantasy destination sign, typical of such events. Rt. 63-Trafford City Express was discontinued on September 17, 1931, long before the advent of PCC cars. The route number, 63, was later assigned to the Corey Avenue route. **(W. D. Volkmer)**

THE SOUTH HILLS LINES

CARSON AND SMITHFIELD STREETS (RAILFAN UTOPIA)

At the city side of the South Hills tunnel, the intersection of Carson and Smithfield Streets offered the railfan photographer a potpourri of subjects. The Pittsburgh and Lake Erie railroad station, the Smithfield Street Bridge, B&O and Pennsy trains, the Monongahela Inclined Plane, and twelve trolley routes provided loads of trolley action for the photographer. Passing this intersection were trolleys on Routes Shannon/Library, Shannon/Drake, 38, 39, 40, 42, 44, 47, 48, 49, and 50. All but the last two mentioned utilized the tunnel to South Hills Junction.

Car 1621 is inbound at Carson and Smithfield Streets having just emerged from the tunnel, on June 29, 1958. (W.D. Volkmer)

One-of-a-kind PCC car 1630 is about to enter the tunnel en route to Mt. Washington on May 16, 1959. The Smithfield Street Bridge over the Monongahela River is at the top of the photo. The P&LE station is on the left and was used by B&O passenger trains operating from Baltimore to points west of Pittsburgh as well as P&LE commuter trains to College, PA. (W.D. Volkmer)

SHANNON-LIBRARY/SHANNON-DRAKE

Following the shortening of the two interurban lines, the Castle Shannon suburban routes still offered some interesting photo opportunities, as a reminiscence of the former interurban operations.

Car 5432 is seen outbound on September 9, 1956 at Washington Junction, where the two former interurban routes diverged. **(W.D. Volkmer)**

PITTSBURGH RAILWAYS COMPANY

Library, Drake and Castle Shannon Trolley Schedules

Effective Monday, October 27, 1958

LIBRARY WEEKDAY SERVICE

Times Leaving Library—To Downtown Pittsburgh:

AM	AM	AM	PM	PM	AM
5:43	7:51	Every 30	3:48	6:59	12:18
6:13	7:59	min. to	4:02	7:39	*12:44
6:37	8:12	PM	4:17	8:19	1.18
6:54	8:29	2:16	4:42	8:59	*1:41
7:02	8:46	2:38	5:05	9:39	2:13
Approx.	9:06	3:00	5:34	10:19	2:43
every 5	9:28	3:19	5:49	10:59	4:13
min. to	9:52	3:35	6:19	11:49	4:48
7:41	10:16				

*To South Hills Junction Only.

Times Leaving Downtown Pittsburgh—To Library:

AM	AM	AM	PM	PM	PM
5:30	8:46	min. to	4:19	6:15	11:20
6:00	9:14	PM	4:22	6:45	AM
6:30	9:30	2:30	Approx.	7:20	12:00
7:00	9:50	2:52	every 5	8:00	12:30
7:24	10:12	3:00	min. to	8:40	1:00
7:41	10:36	3:22	5:29	9:20	1:30
8:00	11:00	3:44	5:38	10:00	2:00
8:20	Every 30	4:04	5:49	10:40	3:30
					4:00

RUNNING TIME

To determine the approximate time trolleys pass the following points add the number of minutes indicated to the times shown leaving the terminal points:

Inbound (Read Down)		Outbound (Read Up)
	Library Loop	41
4	King's School	37
9	Mesta Stop	32
13	Washington Junction	28
16	Castle Shannon	25
21	Overbrook Boulevard	20
30	South Hills Junction	11
41	Grant and Liberty	....

(SEE FOLLOWING PAGES)

In suburban Bethel Township car 1719 is outbound on double track to Library, on Wednesday, August 25, 1954. The line was controlled by block signals along Brightwood Road at Lytle Road stop. The Montour Railroad passed overhead at this point. **(Edward S. Miller)**

Interurban-equipped air car 1617 is on the Drake line along Bethel Church Road, near Brookside stop, on August 25,1954. Cars 1613-1619, and 1645-1648 were equipped for interurban service. They were augmented by twenty-five 1700-series cars used on this route.
(Edward S. Miller)

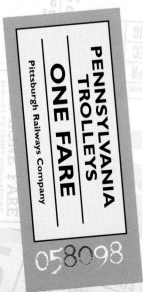

*Since the 1700-series cars were all-electric, they were equipped with inadequate electric horns, and the operator had to rely on this horn along with his bell, his eyesight, and his luck, at these crossings. The operator of 1724 is probably suffering from the notoriously unventilated 1700s, on a hot July 26, 1958 as he crosses a short trestle near the Drake loop. (**W.D. Volkmer**)*

Freshly painted PCC 1633 is inbound on Rt. 48-Arlington over Warrington Avenue. It is about to enter the private right-of-way into South Hills Junction on September 27, 1963.
(James. P. Shuman)

On August 27, 1961 the Pennsylvania Railway Museum Association chartered car 1664, recently redecorated to advertise Mohawk Airlines. There was a short one-block stretch of private right-of-way near the end of the line that was always a photo opportunity during such outings. Mohawk Air Lines later merged into Allegheny Air Lines which later became U.S. Air and more recently U.S. Airways. **(W.D. Volkmer)**

RT. 40-MOUNT WASHINGTON

Specially painted Pittsburgh Pirates advertising car 1633 holds down the Rt. 40-Mt. Washington service at the end of the line, high atop the mountain, at Fingal and Rutledge Streets. The date is August 27, 1961. **(W. D. Volkmer)**

All electric PCC 1735 trundles up Grandview Avenue on June 19, 1964, atop Mt. Washington at Republic Street, where passengers will get a view of the Golden Triangle from above. **(James P. Shuman)**

Much of the former Beltzhoover line survives today as the Arlington-Warrington alternative route to South Hills, if avoidance of the South Hills Tunnel is necessary. During the Pittsburgh Railways era the route climbed the mountain and terminated at the top.

Instruction Car M11 passes a regular service car 1661 on Climax Street near Gearing Street, near the terminus of the Beltzhoover line, on August 10, 1958. The M11, a chartered car, was making its last known run on that date and represented Pittsburgh's first PCC car acquisition. It spent most of its life as a training car for operators. (W.D. Volkmer)

Car 1601 turns off Beltzhoover Avenue onto Warrington Avenue on May 16, 1967, inbound for downtown.
(W.D. Volkmer)

RT. 53-CARRICK
RT. 77/54-CARRICK-NORTH SIDE VIA BLOOMFIELD,
THE "FLYING FRACTION"

Route 77/54 was Pittsburgh's most famous "Flying Fraction." There were other fractionalized routes that resulted from combination of two routes, such as 6/14, 42/38, 34/31, etc., but none were so well revered as 77/54. This route began at Brentwood loop, toured Oakland over Forbes Avenue and Craig Street, and then skirted East Liberty over Penn and Liberty Avenues. It crossed the Allegheny River on the 16th Street Bridge, and looped in Pittsburgh's near North Side business district. It was a belt line in the truest sense of the word.

Car M454 makes a last run cross the 16th Street Bridge of the famous "Flying Fraction" route, Rt. 77/54 carrying railfans on the day of this line's abandonment, September 4, 1965. **(W.D. Volkmer)**

The 1422 is outbound on Route 53-Carrick, September 26, 1963, leaving downtown over the 10th Street bridge over the Monongahela River, en route to the Brentwood loop. Carrick represented one of the heavier patronized lines on the Pittsburgh system. Brentwood was served by no less than three trolley routes, 47, 53, and 77/54. Route 47 was a peak hour only operation. **(W.D. Volkmer Collection)**

THE WEST END LINES

The Golden Triangle 1956.
(W. D. Volkmer)

Pittsburgh's Golden Triangle served as a focal point for the various trolley lines to McKees Rocks, Sewickley, Carnegie, Crafton, and Sheraden. Replacement of the Point Bridge in 1959 spelled the end of these lines.

Car 1797 and an unidentified PCC car pass in the golden triangle area of downtown Pittsburgh with the old Point Bridge in the background. The date is November 22, 1958 and the Rt. 26-West Park car is turning into Penn Ave.
(W. D. Volkmer)

ISLAND AVENUE

First prize for the worst trackage in all of Pittsburgh went to this stretch along Island Avenue in McKees Rocks. The cause of the wiggly rail was the subsidence of the land, as it slowly sank into the Ohio River. Shortly after this April 26, 1959 photo was made of car 1708 (a Washington Chapter NRHS charter), buses had little problem navigating the street. **(W.D. Volkmer)**

The last double-end car line to have a loop constructed allowing PCC cars to operate there was Rt. 28-Heidelberg, an interesting side-of-the-road, single track line in the Borough of Carnegie. Car 1019 was seen on June 14, 1959, operating on that line just before service was discontinued. **(James P. Shuman)**

Car 1795 is inbound on Idlewood Road at the Woodlawn Avenue underpass under the Pennsy Panhandle Division on April 26, 1959. The locale is near the Crafton borough limits, but technically within the City of Pittsburgh. **(James P. Shuman)**

Sheraden-Ingram cars did a jog over the Pennsy's Pittsburgh-St. Louis line also referred to as "The Panhandle Division" on Chartiers Avenue in Corliss. Car 1798 is shown at that location on April 26, 1959, during the last days of the West End lines. **(W.D. Volkmer)**

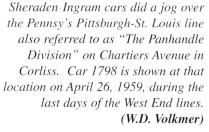

Car 1489 has just left Ingram Carhouse inbound on a road typical of Pittsburgh during the Railways era. This was an example of many places in the city when the city confiscated the trolley right-of-way and forced the railway to pave it, share it and maintain it. **(W.D. Volkmer)**

WEST END CIRCLE

The junction of Crafton Boulevard. (South Main Street) and Carson Street was made complex by the presence of the Pennsy tracks overhead. Inbound Route 26-West Park car 1562 joined routes 27, 28, and 30 from Carnegie at this point, on June 14, 1959. **(W.D. Volkmer)**

McKees Rocks
Rt. 24-Schoenville Shuttle

The McKees Rocks area was home to the main shops of the Pittsburgh and Lake Erie Railroad plus numerous steel mills. It was the focal point for local streetcar riding in the west end.

Route 24 was isolated from the rest of the Railways system in 1919 when the highway department condemned the bridge over the P&LE tracks in McKees Rocks. The bridge was replaced in 1920 without car tracks, something unheard of at the time, especially in Pittsburgh. As a result the line was operated with one lonely car that had to be repaired on line without benefit of a shop, when it needed repairs.

Car 4344 was the last car to ply the line, and when it came due for heavy repairs on May 9, 1952 the line was given over to buses. Here are two views of that car probably taken shortly before the line's discontinuance. (**John D. Seibert, James P. Shuman Collection.**)

Federal Street at Stockton Avenue, Pittsburgh's near north side was thriving as the 1416 passed the 1186, on Tuesday, August 24, 1954. **(Edward S. Miller)**

North Avenue and Sandusky Street was the layover point for Route 22 cars, which looped through downtown Pittsburgh from here. The author has always referred to this location as the Charlie Dengler corner, because Charlie was a railfan/mailman whose route passed by this corner every day. Charlie carried a 116 folding camera in his mail pouch and shot all of the billboard cars during his rounds. The negatives were widely traded around the country. However, Ed Miller shot his own slide of the 1461 here on August 24, 1954. The car was advertising the Urban League at the time. (Edward S. Miller)

Advertising Cars (during the Pittsburgh Railways Company Era)

Many "billboard" cars operated in the pre-Port Authority Transit era, and they could usually be found on the Rt. 22-Crosstown line where maximum exposure could be had to the public.

The 1416 advertised the County Fair, something that at least one car per year did, continuing a long-held Pittsburgh tradition. *(Edward S. Miller)*

Car 1605 advertising Dodge City, a car dealer, is inbound on 6th Street at Duquesne Way on July 19, 1963. *(John F. Bromley, Volkmer Collection)*

Advertising U.S. Savings Bond purchases, car 1463 stops in front of Horne's Department Store on Stanwix Street in downtown on August 20, 1957. *(Peter I. Roehm, Volkmer Collection.)*

A Route 55 car, the 1449, operates east-bound under the George Westinghouse bridge in Wilmerding on September 28, 1963. Automobiles parked too close to the trolley tracks were a nemesis to streetcar operation in Pittsburgh. (*James P. Shuman*)

East Pittsburgh, home to the immense Westinghouse (formerly locomotive building) plant, was served by Routes 55, 62, 64, and 87.

The very steep Library Street hill was a trolley route used only as a diversion for Route 55 cars when the regular trackage became flooded. Fantrip car 5432 posed for these two photos on the hill, in the inbound (downward) direction on September 9, 1956. (*Both, W.D. Volkmer*)

RT. 62-TRAFFORD

The 5.45 mile line from East Pittsburgh to Trafford served, among other places, the Pennsy's Pitcairn enginehouse and classification yards. By 1958, the line was operated by only one weekday car trip, but was seldom missed on fantrips. This often necessitated the fans wielding shovels and brooms to clean out the flangeways if it had rained since the last regular car had passed by.

Very few railfans ever visited Trafford on regular service cars in later years, but it was a "must do" on virtually every fantrip. Car 5432 was no exception, stopping on the hill opposite the still steam-operated Pennsy, on September 9, 1956. **(W.D. Volkmer)**

By April 9, 1961, when fantrip car 1630 posed with the Pitcairn roundhouse in the background, the Railways had already removed all passing sidings on the line, in deference to their redundancy. In May 1962 the line was officially abandoned, apparently without ceremonies of any kind. **(W. D. Volkmer)**

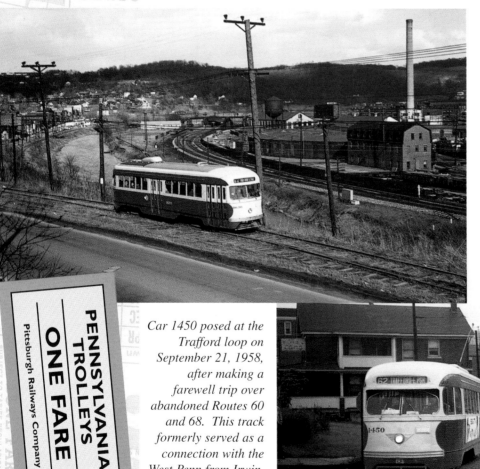

Car 1450 posed at the Trafford loop on September 21, 1958, after making a farewell trip over abandoned Routes 60 and 68. This track formerly served as a connection with the West Penn from Irwin, having been taken out in 1942. **(W. D. Volkmer)**

PENNSYLVANIA TROLLEYS
ONE FARE
Pittsburgh Railways Company

059298

Ardmore Boulevard represented one of the few center of the road private right-of-way operations in all of Pittsburgh. As a result, all fantrips made an obligatory photo stop on Ardmore Boulevard, near the home of one of the fans. (Above) Car 5432 is shown on September 9, 1956, and the 1537 (below) on July 17, 1960 at the same location. **(Both, William D. Volkmer)**

EAST LIBERTY
RT. 75-WILKINSBURG VIA EAST LIBERTY

East Liberty was the first stop on the Pennsylvania Railroad out of downtown Pittsburgh and was served by several trolley routes. These included Routes 60, 71, 73, 75, 76, 77/54, 82, 87, 88, and 96.

The complex intersection of 5th Penn Avenues in East Liberty made for an interesting place to take pictures. Here car 1536 is waiting for the traffic light on Route 75-Wilkinsburg via East Liberty on May 3, 1964. **(W.D. Volkmer)**

Regular Rt. 75 car 1532 lays over near the Pennsy station in Wilkinsburg on Hay Street. It was joined by fantrip PCC 1161 on April 25, 1960. **(W.D. Volkmer)**

RT. 60-HOMESTEAD - EAST LIBERTY

This route was a casualty of the second wave of trolley abandonments after the north side routes in 1957-1959. Routes 60 and 68 were axed on September 20, 1958 due to major reconstruction of Murray Avenue in Squirrel Hill, thus drastically reducing the total railways mileage.

Car 1450 is making a left turn from Shady Avenue onto Penn Avenue in East Liberty on September 21, 1958, during the day-after-abandonment charter run on both lines. **(W.D. Volkmer)**

On the same day, a photo stop is made for the last time on Shady Avenue in Shadyside. This once fashionable neighborhood was a favorite spot for publicity photos of the new PCC cars in the 1930s. **(W.D. Volkmer)**

Bound for Highland Park via East Liberty is car 1506, leaving downtown on 5th Avenue at Ross Street, on August 24, 1954. **(Edward S. Miller)**

Car 1161 is passing the domed arena then under construction, along newly constructed trackage in the median of Centre Avenue, on April 25, 1960. This resulted from a massive urban renewal project in the area. **(W.D. Volkmer)**

WILKINSBURG

The borough of Wilkinsburg was served by Routes 64, 75, 76, 87, and 88 night car up until 1967 when massive abandonments took place.

Car 1433 is turning off of Wood Street onto Penn in downtown Wilkinsburg on May 5, 1965. The Wilkinsburg lines would last until the 1967 massacre of all east end routes.
(Ken Douglas, Bill Volkmer Collection)

Route 76 Hamilton continued to operate until January 1967. Car 1436 makes the turn at Penn and Swissvale Avenues in Wilkinsburg on May 3, 1964.
(W.D. Volkmer)

Car 5432, operating as a charter, stops at the corner of Rebecca and Swissvale Avenues in Wilkinsburg on September 9, 1956.
(W.D. Volkmer)

Homestead - McKeesport
Rt. 56-McKeesport via 2ND Avenue

McKeesport and the Borough of Homestead formed the extreme southeast quadrant of the Pittsburgh Railways Company domain and were steel producing centers nestled along the Monongahela River. Trolleys, of course, carried steel workers to and from their places of work.

The Majestic Mansfield Memorial bridge crossing the Monongahela River provides a backdrop for PCC 1482, as it entered McKeesport on April 24, 1960. Route 68 also served McKeesport but via Kennywood Park. It had been abandoned about eighteen months when this scene was made. Route 99-Glassport, still operated at river level. **(W. D. Volkmer)**

The line featured extensive double tracked side-of-the-road operation along Mifflin Road cross-country en route to McKeesport. Car 5432 was shot moving inbound under the Monongahela Connecting Railroad overpass in that area on September 9, 1956. **(W.D. Volkmer)**

Car 1655 crosses the B&O main on 5th Avenue in downtown McKeesport. Astute railfans kept a sharp eye out for the B&O Baldwin sharks that frequented this location at the time. **(W. D. Volkmer Collection)**

The same crossing is viewed from a different angle, as we see the 1470 coming across the B&O on September 7, 1959. **(W. D. Volkmer)**

RT. 68-MCKEESPORT VIA KENNYWOOD

At 13.8 miles, Route 68 was Pittsburgh's longest streetcar run. (Route 87 Ardmore was a close second, at 13.7 miles). The last day of service for Route 68-McKeesport via Kennywood was Saturday, September 20, 1958. A fleet of new Mack buses took over the next day, so chartered car 1450 had the rails unto itself. The author chose the 1450 for the trip at Craft Avenue Carhouse, because it sported a fresh paint job and with the poor weather that day, it proved to be a wise choice!

The two photos on this page show the curbside loading at a trolley waiting room on 6th Avenue between Sinclair and Locust Streets in McKeesport. West Penn cars formerly shared part of the loop's trackage until their 1938 abandonment in the area. **(W.D. Volkmer photos)**

Pittsburgh Railways served two amusement parks, West View in the north, and Kennywood, in the southeast. A four track layover yard outside Kennywood Park entrance was used by trippers to the park as an alternate destination for Route 55 and 60 cars. Regular service was provided by Route 68-McKeesport, via Homestead, Kennywood and Duquesne. The line was abandoned on September 20, 1958.

Car 1134 (above) is inbound, exiting the long narrow bridge just east of Kennywood Park on June 22, 1958. The view looks east with the park entrance on the left and the loop exit eastbound track in the foreground. By this time, most of the 1000-series PCC cars were in dead storage at Rankin carhouse and many of the 1100s were also there. (Below) Car 5432 lays up in the park during a September 9, 1956 charter excursion. **(Both, W.D. Volkmer)**

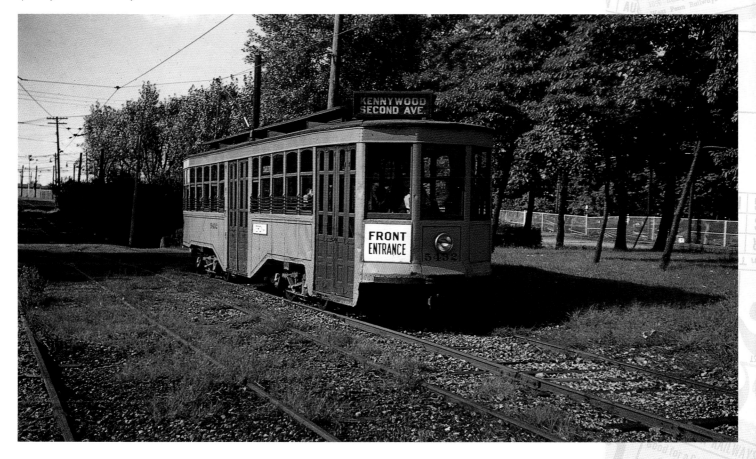

HOMESTEAD
ST. 65 MUNHALL - LINCOLN PLACE

Route 65 was a shuttle operation connecting the Homestead area Routes 55/68 with Route 56, forming the hypotenuse of a triangle. Most of Route 55 was abandoned on July 4, 1964, so for about 14 months the Route 65 cars were extended to cover the outer end of Route 55 to East Pittsburgh. On September 4, 1965, the newly formed Port Authority Transit converted the Homestead area lines to bus. Route 65 was a rather unique operation since it was largely single track with passing sidings, something almost unheard of in the 1960s trolley world. The line was protected by a block signal system. The series of photos on this page depict the last day ceremonies.

Regular Car 1604 leads a parade of buses down 8th Avenue towards Munhall Loop. Most of the sidewalk spectators appear to be only interested in the bus action. Route 65 was one of the few east end routes that did not operate into downtown Pittsburgh. (W.D. Volkmer)

This photo shows the block signals strung along the roadside to protect opposing car movements along Interboro Avenue. Car 1431 was seen there on September 7, 1959. (W.D. Volkmer)

Car 1605 prepares to enter the Lincoln Place loop on the same day, operating south on Interboro Avenue at Mifflin Road. The car was one of the first to be repainted into the gray and white Port Authority colors. (W.D. Volkmer)

*Route 67-Swissvale-Rankin-Braddock traveled the length of the Borough of Swissvale and provided a rather unique trolley riding experience as it traversed a genuine cloverleaf on the approaches to Rankin Bridge, near Rankin car house. Chartered work car M454 met regular car 1756 at Rankin Circle on May 22, 1966. (**W.D. Volkmer**)*

*Car 1059 was seen at Forbes and Shady Avenues on September 21, 1958. (**W.D. Volkmer**)*

OAKLAND

The neighborhood known as Oakland was home to the Pittsburgh Pirates during the days when they played at Forbes Field. It still is home to the University of Pittsburgh and their Panthers. Routes 64, 66, 67, 68, 69, 71, 73, 75, 76, 81 and 77/54 all operated over Forbes and Fifth Avenues giving the area plenty of transit service. Craft Avenue carhouse, one of the city's busiest, served the areas trolley maintenance and storage needs.

Car 1754 makes the turn onto Fifth Avenue leaving downtown Pittsburgh en route to Oakland and Wilkinsburg on June 13, 1959. (**James P. Shuman**)

Forbes Avenue was the main drag in Oakland, where car 5432 is eastbound near Atwood on September 9, 1956. (**W.D. Volkmer**)

The 1513 is inbound on Fifth Avenue approaching the University of Pittsburgh campus on November 15, 1959. (**W.D.Volkmer**)

The West View area was served by a long belt line operation that operated counter-clockwise as Route 10-West View, returning to Pittsburgh as Route 15-Bellevue. Clockwise, Route 15 outbound returned as Route 10. There was also Route 8-Perrysville which paralleled Route 10 on a nearby ridge between downtown and Keating carhouse, near West View.

Car 1286 approached Keating carhouse on a stretch of private right-of-way running on April 23, 1960. **(W.D. Volkmer)**

Car 1299 passes Keating carhouse en route to West View, also on April 23, 1960. **(W.D. Volkmer)**

RT. 21-FINEVIEW

The Fineview line, in addition to offering a "fine view" of Pittsburgh, also represented the steepest grade of any trolley route in the country. As such (12.16%), it was a must for all charter runs, almost without exception. There would ALWAYS be a photo stop at the steepest point on the grade, if for no other reason than to prove that the cars could start on that steep a grade.

Car 1708 is showing her abilities on the grade on an April 26, 1959 outing chartered mainly to tour the west end lines that were about to disappear. **(W.D. Volkmer)**

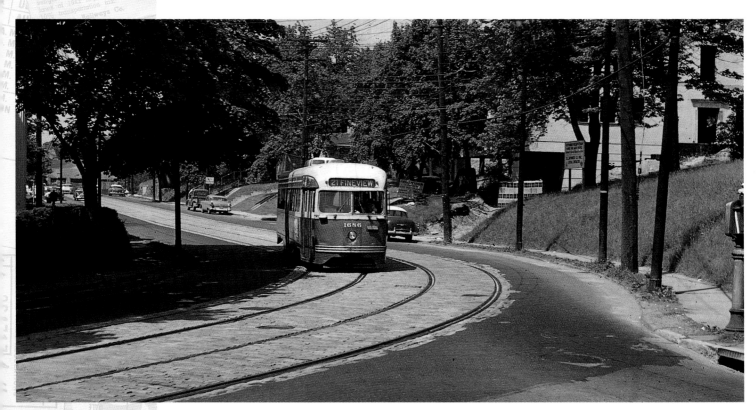

Inbound, the Fineview cars operated over Route 8-Perrysville Avenue offering Route 8 patrons additional inbound service. Car 1686 is inbound on Perrysville Avenue at Federal Street on Friday, May 31, 1957. **(Edward S. Miller)**

FEDERAL AND OHIO — ANOTHER "GRAND" INTERSECTION

Well, OK, it was only a three-quarter grand union, but it offered quite a bit of action, with nine trolley routes passing by. Routes 18 and 19 operated along Ohio Street and Routes 6, 7, 8, 13, 14, 21, and 22 used various segments of the Federal Street trackage. During the 1960s the entire area was razed and a large plaza obliterated the intersection.

Fantrip car 1537 is making the turn from southbound Federal Street onto eastbound Ohio on July 17, 1960. The large building behind the car is the Buhl Planetarium. (W.D. Volkmer)

Car 1090 is eastbound on Ohio Street on June 14, 1959. (W.D. Volkmer)

*One block east of the Federal/Ohio junction was the intersection of Sandusky and Ohio Streets that was almost as complex. Inbound 10 West View car 1788 is on Sandusky Street about to clatter across the special work. Routes 2, 3, 4 and 77/54 joined Route 10 from the right, while routes 18 and 19 came in from the left. Route 77/54 terminated a block or two away. (**Volkmer Collection**)*

SHARPSBURG-ASPINWALL

Route 94-Aspinwall operated over the 62nd Street bridge, through Sharpsburg and terminated in Aspinwall. Prior to 1938, cars of the Allegheny Valley Street Railway, a West Penn affiliate shared the loop at Aspinwall with the Pittsburgh Railways cars. Indeed, for a short while, (about 1910) the West Penn cars operated all the way into downtown Pittsburgh.

*Car 1161 negotiates a passing siding on Freeport Road, just east of Aspinwall on April 25, 1960 during a farewell fantrip to this route. The replacement of the rickety 62nd Street Bridge hastened the line's demise. (**W.D. Volkmer**)*

ROSTER PHOTOS

Car 5445 looks a bit battle scarred, on July 3, 1948 as it rusts out its remaining days in the Tunnel Carhouse Yard. (**Eugene Van Dusen Collection.**)

This concludes our tour of the various trolley lines in Pittsburgh. Next we shall take a better look at the rolling stock by way of roster shots. It is hoped that the areas covered in the roster shots will also fill in some of the locations missed during the tour.

Although low-floor trailer cars were received in Pittsburgh as early as 1910, the motorized versions were built over a ten-year period from 1916 to 1925. There were numerous groups and builders. Space and photo unavailability prevents a more detailed analysis of this very complex roster.

Car 4398 was saved at Glenwood Carhouse for the then fledgling Pennsylvania Trolley Museum since the track connection via the Washington interurban line had been severed. Whenever a fantrip would stop by the carhouse, the venerable double-ender would be fired up for a trip around the carhouse grounds. However, on the date this photo was taken, July 17, 1960, the controller started to smoke at this point behind the carhouse and the run was aborted. The car is alive and well today at Arden. (**W.D. Volkmer**)

CAR 3756

Low floor car 3756 was purchased for the Charleroi and Washington interurban routes, but was later modified with a left-hand door for use on Route 23-Sewickley.

After it was preserved at Arden, it was briefly operated in downtown Pittsburgh to celebrate the Bi-Centennial in 1976. In the photo above, it makes a stop on Wood Street at 4th Avenue and the photo below shows the left side of the car with the extra door, at Fourth and Grant Streets.
(Both, Ken Douglas, Volkmer Collection)

Because the immense PCC car roster of the Pittsburgh Railways contained some subtle differences, a representative photo of each group will be presented here. All PCC cars in Pittsburgh were the products of St. Louis Car Company, St. Louis, MO. Whereas most cities placing large orders of PCC cars split the electrical components 50-50 between GE and Westinghouse, the Railways chose to split 75-25 in favor of Westinghouse for obvious reasons. The Westinghouse cars generally outlasted the GE controlled cars, as they were regarded by most properties around the country as being superior to GE. The converse was ironically true for trolley coaches, of which Pittsburgh Railways owned none.

The M11 was chartered for an excursion (unfortunately its last) on August 10, 1958. In the view above, it was climbing the Henderson Street 12.16% grade on Route 21-Fineview and later (below) crossing a narrow bridge on the Route 13-Emsworth. **(Both, W.D. Volkmer)**

The instruction car M11 was built in July 1936 as car 100 and was taken out of service in late 1958.

PENNSYLVANIA TROLLEYS
ONE FARE
Pittsburgh Railways Company

051118

1000-1099   1937

One hundred cars were delivered between February and May 1937. They were mostly withdrawn from service when some north side and the west end lines were abandoned in the late 1950s.

Car 1013 appeared in a Westinghouse publicity photo (above) when new in the summer of 1937. Shadyside appeared to be a favorite place for such photos. (Below) Car 1019 posed on Fort Duquesne Boulevard along the banks of the Allegheny River in downtown Pittsburgh on June 14, 1959.
(Top, Westinghouse, Bill Volkmer Collection, bottom, W.D. Volkmer)

One hundred more cars were delivered in December 1937 and January 1938. Whereas these cars had the same trolley pole base design as the 1000s, the one-piece destination sign glass was a noticeable change.

Craft Avenue carhouse always presented an impressive lineup of cars with 1110 closest to the camera (below) on August 10,1958. Above, car 1161, one of the last of this series to operate is seen in Aspinwall on April 25, 1960. After abandonment of Route 94, no further use was made of the 1100-series.
(Both, W.D.Volkmer)

1200-1299   1940

Yet another 100 PCC cars were delivered between March and July of 1940. The 1200-series incorporated many design improvements such as forced ventilated traction motors, spring applied, air released tread brakes, elimination of the parking brake, and a larger trolley base for an air intake. None of the 1200-series cars were taken over by Port Authority Transit.

The door side view shows 1242 at South Hills Yard on July 26, 1958 and the blind side view shows car 1257 also at South Hills Yard on June 21, 1958. **(Both, W.D. Volkmer)**

1942   **1400-1499**

Even though World War II was raging in the Spring of 1942, 100 PCC cars were nevertheless delivered to Pittsburgh Railways between February and May of that year. Most of the parts had been fabricated prior to the war's beginning and production had already commenced on the cars.

Car 1406 is operating across the 10th Street Bridge on the 53-Carrick line in June 1963. **(Volkmer Collection)**

Car 1450 is at Kennywood Park on September 21, 1958. **(W.D. Volkmer)**

1500-1564   1944-1945

Wartime production restrictions cut PCC car production by 35%. Only 65 cars were delivered during December 1944 and January 1945.

(Top) Car 1502 at Craft Avenue carhouse on September 7, 1959, (center) the 1540 on Beech Street in East Pittsburgh, and (bottom,) the 1564 at Wood and Penn Streets in Wilkinsburg, September 13, 1964. **(Top and center, W.D. Volkmer, Bottom, Ken Douglas, Volkmer Collection)**

1945   1600

One car 1600 was delivered in September 1945 configured as a prototype all-electric car for the 1700-series cars. The windows on the 1600 could be opened, whereas the 1700s were delivered with sealed windows and forced ventilation. The car was destroyed in a fire at Homewood shops in the early 1950s and we were not able to include a color photo of it. Externally it was similar, but not identical to the 1725-1799-series.

1945   1601-1699

The 99 car 1600-series order was delivered in October and November 1945. Two groups, 1613-1619, (7 cars), and 1645-1648, (4 cars) were equipped with cast-steel pilots and roof-mounted headlights for interurban operation. Several of the cars in this series were renumbered in the 1970s to the high 1700-series (1776 and up) in place of previously scrapped all-electric cars.

Top, Car 1646, one of the interurban 1600s poses on Ardmore Boulevard at Fairfay Avenue in May 1965. **(Matt Herson, Volkmer Collection)**

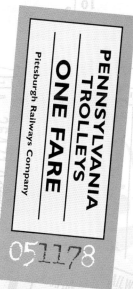

Car 1630 is inbound along Mifflin Road on Route-56 McKeesport via 2nd Avenue on April 9, 1961. This car had just been rebuilt with its unique roof monitor being removed in the process, making it identical in appearance to the other 98 cars in this series. **(W.D. Volkmer)**

1700-1799   1948

The final order of PCC cars and the last streetcars delivered to Pittsburgh Railways were all-electric cars 1700-1799. They were delivered from January to May 1949 (The 1700 was delivered in late 1948). The first 25 cars, 1700-1724 were delivered in the interurban configuration similar to the two groups of 1600s previously furnished. Some of these cars survived in revenue service up until 1998, and several have been preserved in museums.

*Car 1797 is shown (above) sporting the first of the Port Authority Transit paint schemes at 6th Street & Fort Duquesne Boulevard on July 2, 1964. In the center, we see the last car, 1799 at Ingram carhouse on June 22, 1958 alongside one of the then-new GM buses that were taking over. Bottom, car 1799 is waiting to be offloaded on delivery at Millvale Carhouse on May 31, 1949. Cars 1775-1799, being GE motored and controlled did not last well into the Port Authority Transit era as did their Westinghouse counterparts. (**Top and center, W.D. Volkmer, Bottom, John D. Seibert, James P. Shuman Collection.**)*

THE COLORFUL CAR ERA OF PORT AUTHORITY TRANSIT

*The KDKA car, 1777, was renumbered in May, 1973 from 1615 and painted this bright yellow. The photo was made at South Hills carhouse on July 31, 1974. (**Joseph P. Saitta, W.D. Volkmer Collection**)*

During the 1970s Port Authority Transit spruced up the remaining 1700-series PCC cars and several of the better 1600-series for continued operation until money was made available for light rail construction. The disastrous ending of the Sky-bus project in 1975 further delayed the light rail project, making it all the more imperative to keep the PCCs going. The results were a rather striking array of multicolored PCC paint schemes. Lack of space prevents us from illustrating many but we have selected four examples herewith:

*The "Mod-Desire" trolley, 1730, was probably the most rakishly painted trolley ever operated, anywhere. Here it overlooks the city on Route 49-Arlington-Warrington on Sunday October 1, 1972. (**W. D. Volkmer**)*

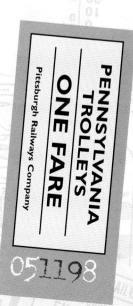

"Don't Leave Home Without Us" screamed the American Express car 1724, turning from Smithfield Street onto Liberty on June 23, 1983. Subway construction had closed Grant Street to streetcars necessitating the Shannon cars to use Smithfield Street instead when making the city loop. **(W.D. Volkmer)**

The "Clark Bar" car was 1742, seen again rounding the corner from Smithfield onto Liberty on June 23, 1983. The subway was finished and put into service on July 7, 1985 ending surface streetcar operation in downtown Pittsburgh forever. **(W.D. Volkmer)**

Car 1630 was built with an experimental forced air ventilation system similar to those in service in Boston at the time. It was a forerunner to the 1700-series cars, but was never duplicated on the air-braked PCC cars in Pittsburgh. The car was chartered by the Pittsburgh Electric Railway Club on March 17, 1957 and made a photo stop at Corliss loop. In the spring of 1961 the car was rebuilt without the ventilator and fans. The roster section of this book includes a photo of the car as rebuilt.
(Richard S. Dickie, Volkmer Collection)

Car 1976 was rebuilt from car 1784 (nee-1603) that had been in an accident resulting in the front of the car being demolished. The windshield on the resulting car was inspired by the then-new Boeing Vertol light-rail vehicle design, and also a PAT desire for larger destination signs. The car is seen here on Grant Street near Liberty on July 4, 1976, holding down an assignment on Route 49-Arlington-Warrington.
(Ken Douglas, Volkmer Collection)

PCC 1734 was out-shopped in 1974 as one of many with unique, colorful paint schemes. On October 13, 1974, it was seen at South Hills displaying a sort of spoof on railfans who were locally dubbed "trolley jollys". **(Ken Douglas, Volkmer Collection)**

Car 1794 was disguised as a riverboat (?) to promote Pittsburgh tourism and was used on Route 49 Arlington-Warrington, a line that gave a panoramic view of the city and river. It was formerly car 1669 and the transformation took place in 1977. Because the car was so tall, it could not be operated through the South Hills Tunnel. Consequently the car was withdrawn from service in 1982 when Route 49 was shut down for reconstruction to light rail standards. **(Ken Douglas, Volkmer Collection)**

Car M210, currently preserved at the Pennsylvania Trolley Museum, is inbound at Carson and Smithfield Streets, on July 22, 1962 to perform some overhead line repairs.
(Robert E. Bruneau, Bill Volkmer Collection)

In 1980 a used sweeper was procured from SEPTA (their car C-145) in Philadelphia and given an extensive rebuilding to add a plow to one end. The car was originally built as a plow and soon converted to sweeper status. In Pittsburgh, it was used briefly on the Shannon line and then given to the Pennsylvania Trolley Museum. In 1997 this car was traded back to SEPTA in exchange for Red Arrow line car 07. **(Joseph P. Saitta, Volkmer Collection.)**

Crane Car M283 is followed by a line truck at 11th Street and Liberty Avenue near Penn Station on August 24, 1954. The red building in the background was later razed to make way for a new Greyhound bus depot. (**Edward S. Miller**)

Twenty years later M283 performs rail replacement chores at the north end of the Mt. Washington tunnel under PAT auspices, on July 7, 1974. Because the tunnel was closed for renovation, the work equipment could be deployed at will for the duration of the project. This car has also been preserved at the Pennsylvania Trolley Museum. (**K.L. Douglas, Volkmer Collection**)

Beginning in 1981 Port Authority Transit undertook a limited rebuilding program to rejuvenate the aging PCC fleet for a few more years until new articulated light rail cars could be procured for operation on the upgraded rail lines. These cars were almost total clones of the PCCs and were built in the company shops using some salvaged (actually very few) PCC parts from 1700-series cars, but making heavy usage of fiberglass for the body work. One of the rebuilds, car 4006, was equipped with air conditioning as an experiment.

Car 4000, the first rebuild, is outbound at South Hills Junction on June 22, 1983. Ultimately 13 of these cars were put into service. It was renumbered 4012 on April 16, 1985 to be similar to the LRV numbers which started at 4101. (**W.D. Volkmer**)

Rebuilt car 4008 is outbound from Castle Shannon on May 2, 1994. This view illustrates the level of line rebuilding, as a result of the advent of light rail in Pittsburgh as outlined in the next section. (**R.P. Townley, Volkmer Collection**)

THE LIGHT RAIL SYSTEM

In the early 1980s, Port Authority Transit dug a subway under downtown Pittsburgh and extensively rebuilt the Route 42/38 Dormont-Mt. Lebanon line. Old Route 38A (Mt. Lebanon-Castle Shannon extension) was also rebuilt, plus remnants of the Library and Drake Lines beyond Castle Shannon and Washington Junction.
A new loop and shop complex was constructed off of the Drake Line at this time allowing demolition of the South Hills Carhouse.

The two views on this page show how the trackage on Broadway in Dormont was transformed for articulated light rail car operation. In the view above car 4105 operates outbound while in the view below car 4141 operates inbound on May 23, 1987, shortly after the line opened in its entirety.
(Both, Joseph P. Saitta, Volkmer Collection)

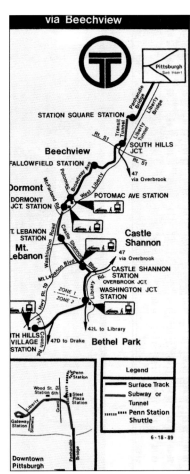

SCRAPPING THE OLD CARS

The three scenes on this page were all made at Ingram Carhouse on August 24, 1954 (about two weeks after the Altoona & Logan Valley system had shut down). These cars were only used in rush hour service on the West End lines, a practice that ended in the spring of 1954. A strike had diminished patronage to the point where PCC cars could handle all runs. A few low-floor cars were stored serviceable. Eventually, only one car, the 5432, was left and it made its final run on September 9, 1956 as profusely illustrated in this book. The 5432 was converted to a paint-striping car M134 and quickly deteriorated to the point where it was not economical to preserve. (*Three photos Edward S. Miller*)

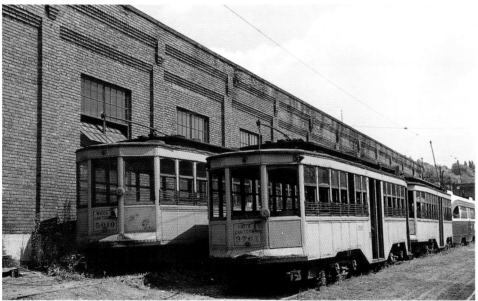

ALTOONA REVISITED

Readers of this volume may wonder why the Altoona and Logan Valley Electric Railway was omitted. The reason is that space limitations forced us to include that city in Volume I of this series. We did, however, receive two nice views of the system after publication and we shall include them here, to round out the western Pennsylvania traction coverage.

We hope you have enjoyed our tour of Pennsylvania Trolleys in color!

Electromobile 72 (Osgood Bradley, 1927) and one other car are shown in downtown Hollidaysburg (above) on the last revenue service date, June 2, 1954. In the scene below three of the five Electromobiles the system owned made one last visit to Eldorado. (**Both, Steve Bogen**)